PRINTING THINGS

Visions and Essentials for 3D Printing

Edited by Claire Warnier, Dries Verbruggen, Sven Ehmann, Robert Klanten

gestalten

Printing Things

Visions and Essentials for 3D Printing

2
→ Theory   19

3
→ Case Studies   55

4
→ Appendix   252

January 28, 2014, marked the seventeenth anniversary of the invention of selective laser sintering, one of the three most common processes in 3D printing. In a lifetime of a human being, a seventeenth birthday wouldn't carry much meaning; you would be approaching adulthood but couldn't enjoy the benefits of it just yet. But in the lifetime of a patented invention, it means you've expired and your inventor-parents are no longer able to keep you chained to the nest. You're free to wander wherever you want, to experiment with dubious materials, and many other creative things your parents never had in mind when you were born.

In 2009 something similar happened to fused deposition modeling or FDM, one of the other major 3D printing processes. A few years earlier, Adrian Bowyer, a senior lecturer in mechanical engineering at the University of Bath, had begun working on an affordable 3D printer which used FDM to print things. He shared the plans for the printer online for others to replicate and innovate upon. A community of enthusiasts, including our own design studio Unfold, kept toiling away, exploring 3D printing on the workbenches, convinced that it would revolutionize the way things are made. It wasn't until 2009, when the FDM patent expired, that a cottage industry of consumer 3D printer manufacturers would emerge and start commercializing open source-based printers. These first 3D printers usually came as plywood building kits that required a small feat of engineering from their early adopters. By 2011 this industry had really taken off to the point where new companies were popping up every week on popular crowd funding platforms, each on a mission to bring 3D printing to the masses. At the same time as the advent of such "do-it-yourself" 3D printers came the first online 3D printing services that did not only target the business-to-business market, as was the case before, but also hobbyists, makers, and creatives. Both events – the increased availability of industrial 3D printers to an online creative community and the development of open-source 3D printers that make the technology accessible – triggered an unprecedented explosion of creative expression and experimentation with 3D printing. Not only as a tool, but also as a medium in its own right.

When carrying out research for this book, we came across a technique that was an early ancestor of 3D printing called photosculpture, dating back to 1860. Here, a revolving series of photographs was reconstructed into a three-dimensional sculpture. As so often happens in these hyperlinked days, we followed the breadcrumbs, looking further into this technology, and our search led us to an article from May 1958 in the *Journal of Photography and Motion Pictures of the George Eastman House*.

George Eastman, the founder of Kodak, developed the first flexible photographic roll film in 1884, replacing the large and heavy glass film plates and boxes of chemicals that a professional photographer would carry around. Six years later, Kodak introduced the Brownie, a very basic and inexpensive cardboard box camera that used the new roll film. Featuring the tagline "You press the button, we do the rest," the Brownie popularized low-cost, hassle-free photography and made it accessible to everyone. While Eastman's company has lost some of its former glory due to the disruption of digital cameras, his democratic vision of photography for everyone is alive and kicking: today's ubiquitous smartphones cameras can even make 3D scans, not too dissimilar to the photo sculpture process.

All of this makes us wonder whether 2009 will turn out to be for 3D printing what the year 1890 was for photography. Considering that photography, cinema and other moving pictures became cornerstones of cultural expression over the course of a century, one must wonder where we're heading with 3D printing and how it will influence our economical, social, and cultural ways of life. We've have only just begun our path down this road...

Claire Warnier and Dries Verbruggen
Unfold

1

→ Practice

What Is 3D Printing?

"3D printing" is the colloquial term for a group of technologies known as additive manufacturing. In order to print a three-dimensional solid object, a 3D printer reads the shape from a digital 3D model file before laying down and bonding successive layers of material. Each layer contains a cross section of the final object, which is formed by stacking these layers on top of each other.

Additive manufacturing is different from most of the traditional manufacturing processes that have existed for centuries because techniques such as milling, sawing, and cutting involve removing or subtracting material from a solid block to obtain to the desired shape. Since a subtractive manufacturing tool always needs a clear path to the area where material has to be removed, there are many restrictions on the types of shapes that can be achieved. The term "subtractive manufacturing" is a recently coined retronym to describe these more traditional methods and to set them apart from the newer paradigm of additive manufacturing. While there have been many "additive" processes in existence throughout history, including bricklaying and welding, these all lack the input of digital information that sets additive manufacturing apart.

3D printing or additive manufacturing is also referred to as rapid prototyping, rapid manufacturing, stereolitography, layer manufacturing, desktop manufacturing, and freeform fabrication. However, the use of such terminology is waning in favor of 3D printing, which has become the most popular term in the mainstream and refers to the low-end range of the additive manufacturing industry.

How Does it Work?

Every 3D print starts with a digital 3D model of the item you want to create. This can be rather something you've downloaded from a sharing site, bought from an online marketplace, captured from an existing physical object using a 3D scan, or modeled from scratch using 3D design software. After the necessary checks to ensure the model is printable or "watertight," a final step is needed to translate the 3D model into a language that the 3D printer will understand. The model is sliced into horizontal layers, and each layer is converted into X and Y coordinates for the printhead. The 3D printer then reads these coordinates and prints the object layer by layer, fusing the material to form a solid physical object.

During the printing process, the partially finished object has to be supported to prevent overhanging geometries or disconnected features in the model from falling over.

Most processes allow you to print in different resolutions. The print resolution describes both the layer thickness and the precision of the outlines of each layer. These layers are always visible, even with a very high resolution, which gives the object a somewhat ridged surface. In many cases, the rough surface of a finished print is smoothed out using post-processing techniques such as sanding, waxing, and polishing.

A

Processes, Materials, and Printers

Additive Manufacturing Process

Over the last 40 years, a wide range of additive manufacturing technologies have been developed with the same goal: to print things. While each of these approaches essentially builds up objects out of individual layers, there is great variety between the different technologies in terms of which material is used, and how the layers are solidified and bonded together. New technologies are being developed all the time, but most can be traced back to just a handful of methods that involve creating objects layer by layer on the basis of a digital file.

Broadly speaking, 3D printing processes fall into two categories. The first group prints objects by using a printhead to bind a pre-spread layer of material, each time tracing and binding the cross sections of the object. The second group uses a printhead to deposit the material itself on the previous section of extruded material.

Today, the most widely used 3D printing processes are the trio of SLS (selective laser sintering), FDM (fused deposition modeling), and SLA (stereolithograhpy).

Unfortunately, there are as many terms and abbreviations to describe the various processes as there are manufacturers and printer models. For this overview, the most common and well-known terms are used to describe each process. Where necessary, alternative terms for the same process are supplied together with the more generic industry standard term.

A.1 Binding Processes

Binding processes work by spreading a complete layer of a powdered, liquid, or sheet material across the entire build volume before a tool head draws and binds the contours of the object. An entire layer of material is deposited each cycle, so the printed object is supported by unused build material at all times, which mitigates the requirement of extra generated support structures – except in the case of stereolithography.

A.1.1 Stereolithography (SLA)

Stereolithography (SLA) was the first 3D printing process to make the leap from lab experiments to commercialization. Stereolithography is a process that uses photopolymerization, a technique which was also used in the traditional paper printing industry to produce embossed flexographic printing plates. A photopolymer is a liquid polymer, a resin that solidifies when exposed to light. A laser draws the sections of the object on the surface of a bath of a liquid photopolymer resin, solidifying the cross section. The first cross section is built on top of a platform, which is immediately submerged below the surface and lowers a fraction after each completed layer.

SLA is very suitable for printing highly detailed smooth parts at a reasonable speed and is mainly used for "look and feel" prototypes. The popularity of SLA in the early years of additive manufacturing explains why it is commonly referred to as rapid prototyping.

A variety of resins are available with different properties resembling those of common engineering plastics, including rubber-like materials. Color ranges are very limited, and many resins are transparent or translucent for optimal curing using light exposure. There are also opaque resins, but these are typically limited to black, gray, white, or brownish tints.

Unlike other binding processes, there is no support material inherent to this technique. Although the resin bath gives some amount of support to the printed layers, it is too liquid. This means that large overhangs and disconnected parts need support structures. Such structures are always produced using the same material and are of the break-away type because the SLA process does not allow for multi-material printing.

SLA was developed by Chuck Hull in the mid 1980s and was commercialized by 3D Systems, Inc., which he founded in 1986. In 2012 FormLabs introduced the first desktop-class 3D printer based on a stereolithographic process, the Form 1. The industry standard term for stereolithography is vat photopolymerization.

Processes, Materials, and Printers	Binding Processes	A.1.2

A.1.2
Selective Laser Sintering (SLS)

Selective laser sintering is another major 3D printing process, and one in which a strong laser melts and bonds a powdered material together by heat. A thin layer of powdered material is deposited by a roller on the build platform, after which the laser draws the sections of the object on the powder, transforming it into a solid material. After each laser pass, the platform lowers and a new layer of powder is spread over the top of the object from a container on the side.

SLS is a relatively fast process that is well suited for printing structural parts and pieces that show limited layering. It is therefore a popular process for end-use manufacturing.

Creating objects by fusing powder has proven to be a very successful and adaptable 3D printing process. The machines from the main vendors mostly use plastic powders, especially nylon, which is strong and fairly flexible. However, almost any material that can be supplied in powder form and that melts under intense laser heat can potentially be used as 3D print medium. Many different processes have been developed based on SLS over the years, extending the field of application to composite materials, metals, ceramics, glass, and sand. These processes have sometimes adopted different names, such as direct metal laser sintering and selective laser melting. In most laser sintering processes, the materials used retain their natural color, but the nylon often used in the production of consumer goods can easily be treated using the same dyes used for textiles.

Support material is inherent to the process, so there is no need to add, or later remove, support structures. The object being printed is embedded in unfused powder, which also serves as a support structure. Because it is a powder, it can easily be removed using a vacuum cleaner and pressurized air, even from tiny crevices. This is a great advantage of powder-based processes, and one that means it offers the most design freedom of all processes. Any excess unfused powder is removed and to some extent recycled in the next print job.

SLS was developed by undergraduate student Carl Deckard and assistant professor Joe Beaman at the University of Texas in the mid 1980s. In 1988 both men founded DTM Corp. (a reference to desktop manufacturing), which went on to commercialize SLS. In 2001 DTM was bought by 3D Systems, inventor of the SLA process. Another large manufacturer of SLS 3D printers for plastics, metals, and other materials is German-based EOS. In early 2014 the original SLS patent expired, paving the way for more accessible SLS-based 3D printers. The industry standard term for selective laser sintering is powder bed fusion.

Processes, Materials, and Printers	Binding Processes	A.1.3

A.1.3
Inkjet Powder Printing (3DP)

Inkjet powder printing uses a similar powder bed approach to SLS. Instead of fusing the powder with a laser, however, a binder is sprayed over the material to glue the particles together. This process uses inkjet printheads that are very similar to those found in your desktop printer, and the process therefore looks very similar to printing on paper. Because of the use of an inkjet printhead, some inkjet powder printing machines offer full color printing. Parts produced by 3DP are more fragile than parts produced in processes that melt and fuse material. In a post processing step, the fragile objects are infused with resin to make them stronger and to enrich the color saturation.

3DP is mostly used as a cost-effective way of producing visual prototypes. Its biggest selling point is full color printing: this is a capability mostly unrivaled by other processes.

Most inkjet 3D printers use a gypsum-like powder which hardens when sprayed with a liquid binder. By adding multiple inkjet printheads to the machine, each filled with binder of a different color, inkjet 3D printers are able to produce full color objects, mimicking the visual appearance of other materials. This makes the technique ideal for the photorealistic reproduction of portraits or figurines. Fairly recently, people have started experimenting with the use of alternative powder compositions in standard inkjet powder printing machines in order to print ceramics and glass, for example. In such cases, clay or glass powder is temporarily glued together and later sintered in a kiln – a process in which the binder burns out. Professor Mark Ganter from the University of Washington maintains the Open3DP blog, which documents open-source recipes for a wide range of powdered materials and binders.

As with other binder-based processes, support material is inherent to 3DP, so there is no need for separately generated support structures.

3DP was developed at the Massachusetts Institute of Technology (MIT) in the early 1990s by Michael Cima and Emanuel Sachs, and later licensed to Z Corporation, which is now part of 3D Systems. The officially registered term for 3DP is "three-dimensional printing," but in order to not confuse this with 3D printing – a term that now describes the whole field – inkjet powder printing is used here for this specific process. The industry standard term for inkjet powder printing is binder jetting.

A.1.4
Laminated Object Manufacturing (LOM)

Laminated object manufacturing printers produce objects by laminating thin sheets of material onto each other. The cross sections are cut from the sheets with a knife or laser cutter. When the print is finished, the model needs to be removed from the solid block of material into which it is embedded.

LOM is in limited use. However, in the case of paper (the most popular sheet material in use), the build material is low-cost and readily available. The final material has a nice wood-like quality to touch, and recent improvements to LOM allow for full color printing using a similar inkjet approach to 3DP.

Given the use of a solid build material, support material is inherent to the process and there is no need for separately generated support structures. Excess material is crosshatched to facilitate easier removal, but it is difficult if not impossible to remove from hollow spaces and small crevices.

LOM was developed by California-based Helisys, Inc. in the late 1980s. Many companies that offered LOM 3D printers went out of business, but Mcor Technologies, an Irish company, is reviving the technology with the addition of full color 3D printing and other improvements. Staples, a chain of office supply stores, launched a pilot program in Belgium and the Netherlands in 2013 in partnership with Mcor to offer a paper-based 3D printing service to its customers. The industry standard term for laminated object manufacturing is sheet lamination.

A.2
Deposition Processes

Deposition processes extrude a liquefied material through a nozzle in the printhead, which deposits the material in a line pattern, layer by layer, on a lowering build platform. In contrast to the family of binding processes, the model is not embedded in the material, but stands freely on the build platform. This means that additional support structures for overhanging geometries must be created.

A.2.1
Fused Deposition Modeling (FDM)

Fused deposition modeling is a 3D printing process in which a thermoplastic filament is fed into a melting chamber and subsequently extruded through a nozzle. The object is drawn with these melted threads, which bind to the previous layer by heat and harden soon after extrusion. The process works in a similar way to a hot melt glue gun, which is incidentally what inspired the invention of FDM.

Because of the use of standard thermoplastics, fused deposition modeling is well suited for the production of functional parts with great mechanical strength, but the surface shows more layering than other techniques.

Fused deposition modeling works mostly with thermoplastics such as nylon, ABS, PLA (a biodegradable plastic), and polycarbonate. Experimental composite materials have been developed which combine a thermoplastic binder material with a powdered filler material such as wood, metal, or ceramic. Additionally, experimental systems have been developed which directly melt certain metals.

FDM is generally limited to printing in one solid color. Most high-end machines have two printheads, but one of these is always used for soluble support material. Some desktop 3D printers also come with two heads, and these support printing with two colors – just imagine a blue globe with green continents.

There is no support material inherent to this technique as with selective laser sintering, so support has to be generated and printed. These structures may be break-away support, which is printed in the same material as the final object and removed using tools. Support structures can also be printed using a soluble thermoplastic material from a second printhead, which is dissolved in a special chemical bath that leaves the material of the final model intact. Most desktop 3D printers have only one printhead, and in this case the need for support is often mitigated by designing around the problem.

FDM was developed by Scott Crump in the late 1980s and commercialized by Stratasys, the company he founded in 1988. Some of the key patents on FDM expired in 2009, leading to a surge in open-source alternatives like the RepRap Project, out of which most current desktop 3D printers developed, for instance the MakerBot (MakerBot is now a subsidiary of Stratasys) and the Ultimaker.

The term fused deposition modeling and its abbreviation FDM are trademarked by Stratasys Inc. The equivalent term, fused filament fabrication, or FFF, has been adopted by the RepRap community as a legally unconstrained alternative. Some manufacturers use the term plastic jet printing, while the industry standard term for fused deposition modeling is material extrusion.

| Processes, Materials, and Printers | Deposition Processes | A.2.2 |

A.2.2
Paste Extrusion

Paste extrusion is a process related to FDM in which a material is extruded through a nozzle and deposited in a long thread that "draws" the object. Instead of melting a plastic filament wire, paste extrusion, as its name implies, uses a cold or moderately heated material with a paste-like consistency. The material is then cured or dried by air.

Paste extrusion is mostly used in experimental or open-source 3D printers in order to experiment with a wide range of materials in 3D printing applications.

Given the low technical requirements of working with a paste extrusion process, almost any material that can be squeezed out of a nozzle can be used as a print medium. Common paste extrusion materials are concrete and cement (for the production of large-scale architectural structures), clay (for intricate ceramic pieces), precious metal clay, various foods like chocolate or dough, and biomaterials for the experimental production of living tissue.

Most paste extrusion 3D printers have been developed as part of research projects, and the process is also popular in the open-source 3D printing community (the Fab@Home 3D printer has a paste extruder). In addition to its low technical requirements, this is due to its versatility and ability to be combined with more craft-like processes. Together with FDM, paste extrusion falls under the industry standard term material extrusion.

| Processes, Materials, and Printers | Deposition Processes | A.2.3 |

A.2.3
Polyjet

The polyjet process is one of the most recent additions to the family of 3D printing processes. Like SLA, it uses a liquid photopolymer build material. Unlike SLA, however, where the object is submerged in a bath of resin and solidified layer by layer using laser light, polyjet uses a printhead similar to that of an ink-jet printer to directly jet or deposit micron-scale droplets of photopolymer on the object. As soon as each layer is deposited, the still-liquid top layer is cured with UV light.

Polyjet is a high-definition, multi-material, multi-color process that is well suited for producing the most realistic prototypes with the smooth look and feel of a finished product.

By utilizing different printheads and cartridges – just like a computer printer – the polyjet process is able to simultaneously print multiple materials and colors, coming very close to a universal process capable of printing multi-color, multi-material objects. The most high-performance machines can produce up to 14 materials and colors at the same time from a spectrum of more than 100 options. This means that the material properties of the same printed object can vary from rigid to flexible and from colorful to transparent.

Support structures are needed in the polyjet process, and these are printed by one of the printheads in a gel-like material that can be washed away once the print job is finished.

Polyjet was developed by the company Objet in 2000 and is marketed under the term polyjet. In 2012 Objet merged with Stratasys, makers of FDM machines. The industry standard term for polyjet is material jetting.

Personal
3D Printing

B

B.1
Online 3D
Print Services

Professional 3D print bureaus that offer their services to businesses have existed since the birth of 3D printing. However, the current popularity of 3D printing can be attributed in part to a handful of innovative 3D print service bureaus which, at the end of the last decade, recognized the potential of opening up professional industry-class additive manufacturing to individual designers, artists, engineers, and other makers. These companies offer easy-to-use online services that provide immediate quotations for uploaded 3D files in whatever material is specified. More important, their services are paired with a thriving online community which supplies tutorials and support for novice creators.

Popular print services are:
Shapeways: www.shapeways.com
iMaterialise: www.imaterialise.com
Ponoko: www.ponoko.com
Sculpteo: www.sculpteo.com

Most of the services also function as an online market place. They offer an Internet storefront where designers and creators can host their own virtual products. The moment someone orders one of the designs on offer, the print service will print it on demand and ship it to the consumer, paying a percentage fee to the designer. This benefits both the designer and the print service – the latter is able to source content for its service while the former can create a buying public for his work without having to invest in physical products or create an inventory.

B.2
Local 3D Print Shops

In more and more cities around the world, 3D print shops are starting to pop up, offering 3D printing and assistance to the local community. They range from a more do-it-yourself approach with desktop 3D printers available for public use to more full-service shops that employ one or more professional machines. These 3D print shops sell both objects and print services and have the advantage of visibility on the high street and personal customer care.

Another local option when realizing digital designs is a fab lab, which stands for fabrication laboratory. A fab lab is a small-scale workshop in which you are encouraged to use 3D printers and other digital manufacturing tools yourself, with assistance provided by volunteers and other users. The concept of fab labs was initiated at the MIT Media Lab at the Massachusetts Institute of Technology. Today, fab labs can be found in almost every major city in Europe and the US. To find a fab lab in your neighborhood, you can consult the list on the official MIT Fab Lab web site or the Fabwiki.

Locate a fat lab near you:
MIT Fat Lab: fab.cba.mit.edu/about/labs
Fabwiki: wiki.fablab.is/wiki/Portal:Labs

Find a local print service:
3D Hubs: www.3Dhubs.com
Makexyz: www.makexyz.com

Recently a few initiatives have sprung up which link an online service platform with local 3D printing facilities. Anyone can advertise their print services on the platform, from people with a 3D printer at home to local service outlets. Designers and makers looking to have a part printed can use the service to find a local print service that meets their budget and other requirements. Examples of such platforms include 3D Hubs and Makexyz.

B.3
Customization Apps for 3D Printing

With the demand for 3D-printed consumer goods rising, so is the demand for personalized goods. There are plenty of customization tools available online that can be used to create personal 3D-printed objects such as personal figurines, lampshades, smartphone cases, and jewelry. These customization apps are often hooked up to one of the major print services (e.g. Shapeways, Sculpteo, or i.materialise). Some of these tools are offered by the service providers themselves, mostly to demonstrate their capabilities, but more and more designers are also offering their products in combination with a customization app with the help of API and programming tools that facilitate connection with a print service.

Selected online 3D print services:
Shapeways: www.shapeways.com/creator
i.materialise: www. imaterialise.com/creation-corner
Sculpteo: www.sculpteo.com/en/workshops
Thingiverse: www.thingiverse.com/apps/customizer

Selected companies and designers that focus on customizable design and products:
Twikit: www.twikit.com
Mixee Labs: www.mixeelabs.com
MWOO: www.mwoo.me
Society for Printable Geography: www.printablegeography.com
Nervous System: www.n-e-r-v-o-u-s.com/tools

The various online 3D print services are a good place to start looking for apps that allow customization and personalization. Each of these has a dedicated page showcasing the apps that can be used with their services.

The market for customization apps is still in its infancy, but there are already some pioneering companies and designers who focus on customizable design and products.

B.4
3D Modeling Applications

3D design software used to be the domain of specialized professionals. With steep learning curves and high pricing, these packages were beyond the reach of many creatives and makers. The rise of consumer 3D printing brought with it user-friendly applications, often free, geared towards 3D printing. Most of these communicate directly with the most popular 3D print service providers. Alternatively, users can download designs as STL files, the de facto standard for 3D printing.

These applications can be divided into a few groups, each of which has a different way of creating 3D objects.

"Building-block" approach applications:
123D Design: www.123dapp.com/design
Tinkercad: www.tinkercad.com
3D Tin: www.3dtin.com

One category is the building-block approach. Applications in this group use standard shapes that can be modified and combined to create an object. In 123D Design, users can employ standard shapes like cubes, spheres, and cones, as well as more complex forms, which might include parts for robots or bicycles. Tinkercad and 3D Tin follow a similar approach.

"Task like a sculptor" applications:
123D Sculpt: www.123dapp.com/sculpt
Sculptris: www.pixologic.com/sculptris
Leopoly: www.leopoly.com

Another method of shaping a digital form is to approach the task like a sculptor. Here, users start out with a large block and can create new shapes by pulling and pushing the outline. Applications like 123D Sculpt, Sculptris, and Leopoly use this sculptural method of creating objects.

"Creation of figurines" applications:
123D Creature: www.123dapp.com/creature
Skimlab: www.skimlab.com
Archipelis: www.archipelis.com

A third group of applications focus on the creation of figurines. 123D Creature, Skimlab, and Archipelis are applications that enable users to create puppets and figurines. 123D Creature and Skimlab combine the sculptural approach with a skeleton, making it possible to mimic a variety of poses for your creature. The Archipelis interface is more similar to sketching a design.

A very popular application these days:
Printcraft: www.printcraft.org

One very popular application these days is Printcraft, the print application for the successful game *Minecraft*. The objects and figurines created in this game can then be printed thanks to this application.

An application for a more technical approach:
OpenSCAD: www.openscad.org

OpenSCAD is a free solid 3D CAD modeler for those who are not daunted by a steeper learning curve and who want to create technical parts. It is mostly used by engineers and programmers. The software does not focus on the artistic aspects of 3D modeling, but technical aspects for the creation of machine parts.

Two more free 3D applications:
SketchUp: www.sketchup.com
Blender: www.blender.org

Two more free 3D drawing applications are SketchUp and Blender, but these were intended more for visualizing ideas, animations, and architectural and interior drawings rather than 3D printing. That being said, SketchUp has lately focused more on the 3D printing aspects of the modeling software.

An application that can be used on a smartphone:
123D Catch: www.123dapp.com/catch

One final application of interest from the Autodesk suite of consumer apps is 123D Catch, a 3D scanning application that can be used on a smartphone.

B.5
File-sharing for 3D Printing

Dedicated design repositories and community platforms have been established for people searching for printable objects for their 3D printer and digital designs that can be modified. On the other hand, users can also share their own designs with the world. The best known of these platforms is thingiverse.com, the 3D file sharing community from MakerBot, manufacturer of consumer 3D printers.

All the content is free to use, although the rights for some objects are restricted to non-commercial use. A similar platform is youmagine.com, which is operated by Ultimaker, another 3D printer manufacturer.

The Pirate Bay, a web site that facilitates peer-to-peer file sharing using the Bit-Torrent protocol (and that is notorious for facilitating illegal downloads of copy-right content) has recently added a category for 3D-printable files: Physibles.

Another community for sharing material and finding inspiration from other users is instructables.com. Not only 3D-printable designs are shared here, but all kinds of manuals and instructions on how to make and create a vast range of items, from coatracks to biogas installations – and even a guide on how to cut onions.

Personal 3D Printing   B.6

B.6
Consumer
3D Printers

If you want to own your own 3D printer, it is now possible to do so without spending a fortune. The first efforts to develop open-source 3D printer projects were kick-started around 2006, and things have come a long way since then. Today, there is an abundant choice of models, ranging from €250 DIY kits like Printrbot to fully assembled prosumer machines such as the Ultimaker 2, which retails for a little under €2,000. Almost all consumer 3D printers have evolved from the open-source RepRap printer and use the FDM technique to meld ther-moplastics like ABS or PLA.

Although consumer 3D printers are not as advanced as their industrial counter-parts and require more tweaking, adjustment, and care, there are many advan-tages of having your own unit at home or at work. With the exception of some top models, most consumer printers come as building kits. It takes some time to assemble them, but this is a process that gives users great insight into the 3D printer as a machine and how it works. These printers are not "black boxes" like their larger industrial siblings: users can understand the operation mechanism without too much difficulty as they become familiar with the ins-and-outs of the process. This also means that the printers are fairly easy to repair. Indeed, repair work is almost an inevitability, but fret not: most manufacturers have excellent communities ready to help. Moreover, given that these printers have an open-source design and users have the experience of putting them together them-selves, it is encouraged to hack and modify the system in creative new ways.

The market for consumer 3D printers is now maturing to the point that we see more and more manufacturers making the transition from DIY kits to fully as-sembled machines with user-friendly software and services. Moreover, the big names in the industry, Stratasys and 3D Systems, have also made large forays into the consumer market. Although users will be less able to repair or hack pre-assembled printers and have to forgo the insights gained by building such de-vices, there is also an obvious benefit. Ready-to-use printers offer greater con-venience, reliability, and hassle-free printing. The company supplying the printer also offers support, software, and specific materials for the unit. For those who want to dive straight into printing things, they offer a good experience.

If you are in the market to buy a hobby, consumer, or prosumer 3D printer, a good place to start is the annual MAKE Ultimate Guide to 3D Printing. This publi-cation features extensive expert reviews of the most popular models of at-home 3D printers.

2

→ Theory

A

Empowerment

→ #authorship   → #empowerment   → #open source

Adrian Bowyer, *RepRap*

The Internet gives people power. The power to communicate, to overcome physical distance, and to share ideas. When connected to the Internet, digital fabrication methods can convert this power into tangible objects. If you have a computer that is connected to the Internet and a 3D printer, the world is at your feet. It is a story we hear very often nowadays – a story about involvement, DIY enterprise, and grassroots projects. But how liberating is this empowerment? And are the Internet and 3D printing really shifting the power of production into the hands of the masses? Is this phenomenon set to finally fulfill the ultimate dream of Karl Marx?

Knowledge Sharing

In 2005 Adrian Bowyer from the University of Bath in the UK began research on a machine that could reproduce itself. He focused on the development of a simple, open-source 3D printer. Some parts of this printer could be found in a local hardware store, while more complex parts could be 3D-printed. The first mother machine he developed with parts printed on a commercial 3D printer was then able to print the parts for its own "offspring" – and thus reproduce itself. In 2006 the first Rep-Rap ("replicating rapid prototyper") went on to print the parts for its first child.

To promote his idea of a self-reproducing machine, Bowyer handed over his plans, technical specifications, and software to the public domain as open-source files that could be shared and improved on by the online community. The only thing Bowyer asked for in return was that people let any machines they built first print the parts of a new machine so that it could replicate itself at an exponential rate.

As a readily accessible open-source tool, the population of RepRaps was steadily growing and improved rapidly with each successive generation. The project became very popular. Soon after the first RepRap saw daylight, building kits containing all the necessary parts started to appear on the market, making it more convenient for people to assemble their very own 3D printer. Small companies also started to develop other printers, based on the RepRap, that were cheaper, faster, or more user friendly. Projects such as W.Afate started to emerge, enabling less wealthy countries to create their own version of the RepRap 3D printer. Afate Gnikou and WoeLab from Togo applied RepRap technology to the growing problem in Africa of electronic waste. By using waste products as a raw material and transforming old computer and electronic parts into 3D printers, a problem was cleverly transformed into an opportunity.

The RepRap and its derivatives brought people into contact with a technique that had previously been very expensive, specialized, and therefore inaccessible. Designers, artists, and makers could now invest in a machine for a couple of hundred dollars rather than tens of thousands of dollars. RepRap and its spiritual father Adrian Bowyer were the spark that ignited the current explosion of affordable 3D printers for home and small business use. While it is impossible to predict the often unexpected changes brought about by such disrupting technology, the question remains as to whether at-home 3D printers will become mainstream and transform every household into a small production unit. This seems rather unlikely at present. Nevertheless, current technology has brought the idea of hands-on 3D printing from the realm of science fiction, making it a reality. This development alone has inspired many novel and creative applications, and this book presents some of the results of a process that is more accessible and adaptable.

A couple of years before the RepRap made its appearance, Neil Gershenfeld from the Massachusetts Institute of Technology was working on a system to make highly specialized digital manufacturing tools accessible to a large group of people. In 2001 he realized the first fab lab at the MIT. The MIT Fab Lab (short for both "fabrication laboratory" and "fabulous laboratory") contained highly specialized digital production tools such as 3D printers, CNC milling machines, and laser cutters, allowing students to make whatever they could imagine. Gershenfeld had noticed that his class entitled "How to make almost everything" was immensely popular with students. Although the original target audience was engineers, many architects, designers, artists, sociologists, and others applied for the course. Students were thrilled by the title and the prospect of working hands-on with advanced production tools. After these tools were made accessible, new projects and ideas quickly developed. Students started to work together, pooling their skills and knowledge to achieve their goals.

It was an eye-opener for Gershenfeld to see how knowledge and technology could be transformed into ideas and projects just by opening the arena to different kinds of people, and so the first fab lab was born. Nowadays, there is a worldwide network of about 300 official fab labs in operation, and the number is still growing. They can be found in major cities around the developed world as well as in remote locations in developing countries and even in war-torn regions like Afghanistan. Fab labs are open for everyone to use, and in every location the sharing of knowledge and ideas is the key philosophy behind their existence. People who want to create their project in a fab lab can work there for free, or pay-

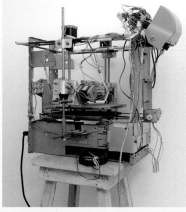

Afate Gnikou and WOELab

ing for their expenses, as long as they share their knowledge with others in the fab lab and the rest of the world through the public network. If someone is not willing to share their project, they may still work at the facility but are required to pay a commercial price for using the machines.

Bridging Physical Distances

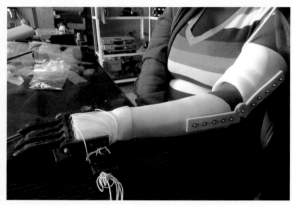

Richard Van As and Ivan Owen, *Robohand* → **p. 201**

Knowledge sharing is one of the main features of an open source community, and it is one of the key reasons why open source has become such a successful model. Although sharing ideas can prove difficult and problematic with issues such as copyright and profiteering being inevitable, it is also inspiring to see how beneficial it can be to engage people in collaborative projects on an unofficial and non-commercial basis.

The Robohand by Richard Van As and Ivan Owen is an energizing example of how the Internet can connect people with different backgrounds and knowledge, enabling them to create and to motivate a whole community to work towards a common goal. Richard Van As lost the fingers from his right hand in a tragic accident in 2011. He later realized that while prostheses for hands are quite common, there are fewer options available for lost digits. Furthermore, the solutions that are available are very expensive. Against this background, he decided to find a way to create an open-source, low-cost prosthesis for his lost digits. While surfing the Internet in 2013, he found a YouTube film by a mechanical engineer named Ivan Owen, who was demonstrating a mechani-

cal hand he produced for fun. Van As was exited about the project and contacted Owen to ask if they could work together on a mechanical finger for his disabled hand. Not hampered by living on opposite sides of the world, the two men agreed to embrace the challenge together. Owen designed the parts using CAD and emailed the digital files to Van As, who then produced the parts. After assembling everything and testing the device in person, he emailed some sketches with improvements back to Owen for another iteration of the design. Other people noticed the project, and soon the Robohand project started to flourish.

By using the Internet as a communication medium, blogging every step of the development, the design team found many more people who wanted to be involved. MakerBot, a producer of at-home 3D printers born out of the RepRap project, donated a couple of their desktop 3D printers, and soon universities and specialists were freely offering their assistance and knowledge to make the Robohand better and safer. The design is still being improved while helping more people with missing arms, hands, and digits. Naturally, caution is needed if people use online resources create their own medical care, but this approach can also open up specialized care to vast populations who never had access to it in the first place.

The design files for the Robohand can be found as a free download online on Thingiverse, producible on any 3D printer of choice. 3D repositories like Thingiverse and YouMagine are community web sites that allow everyone to share their ideas and digital product designs with others. The 3D-printable products found there range from gadgets to technical parts (including many 3D printer improvements) and open-source versions of traditional products such as Leo Marius's Open Reflex camera. After downloading a design from Thingiverse or YouMagine, users can produce it, take a picture of it, and are also encouraged to share their experiences with the object as well as any suggested improvements with the community.

Shared Knowledge Brings Shared Responsibilities

The Liberator 3D-printed gun from Defense Distributed brought the issue of responsibility to the media's attention. Although the printed gun is more a political statement than an actual weapon – and you had better not fire it because it kills from both ends! – the product triggered a mediated discussion on the ethical questions of where 3D printing, open source, and shared knowledge can lead. It also inspired a witch hunt of its own in which police officers confiscated a 3D printer and some 3D-printed parts because the parts, meant to improve the printer itself, were incorrectly identified as parts for a printed gun. The digital files of the gun were distributed online via The Pirate Bay. Although The Pirate Bay itself is against weapons, the site did not take the files offline, stating: "We believe that the world needs less guns, not more of them." In any case, once these files had been uploaded to the Internet, it would be impossible to regulate their distribution. "These prints will stay on the Internet regardless of blocks and censorship, since that's how the Internet works." True to the philosophy of The Pirate Bay, the final argument is a plea for accessibility, even in this extreme case: "Better to have these prints out in the open Internet and up for peer review (the comment threads), than semi-hidden in the darker parts of the Internet."

The public outrage surrounding 3D-printed weapons was too visible to ignore, but its overexposure in the media and the subsequent misinformed reactions obscured the more interesting discussions. Anecdotal in this context is a story from the renowned Victoria and Albert Museum in London. When they acquired one of the first prototypes of the USA-made printed gun for a design exhibition entitled "Design is Everywhere," it was held up at customs and did not arrive in time, even though the museum is a licensed weapons importer (the collection includes many historical weapons). Notwithstanding the immense bureaucratic effort to acquire an "original" prototype, the real prototype was all the time available online as a digital blueprint. A 3D print studio in London was able to produce a copy for the V&A just in time for the exhibition opening. This is most likely the real power that shared knowledge and 3D printing will bring.

Defense Distributed, *The Liberator*

B

The Right to Copy?

→ #authorship → #customization

As you can see by flicking through the pages of this book, a large proportion of 3D-printed output is truly original content. However, given the increasing democratization of manufacturing and the subsequent influx of consumers turned casual product designers, another large share borrows aspects small and large from already existing things, be they digital or physical. Such borrowed elements may infringe on third-party intellectual property, including copyrights, patents, and trademarks. The situation is not too dissimilar from the disruption caused by the invention of the printing press in 1450 or the MP3 music file format together with online file sharing in more recent years. In short, 3D printing is challenging our existing paradigms of authorship, originality, and property.

A Tale of Two Copies

In July 2013, director and 3D animator Joaquín Baldwin sent a set of colorful game figurines to Shapeways, a popular 3D printing service. The 3D models were taken from *Final Fantasy VII*, a game from the late 90s that triggers waves of nostalgia in those who spent countless hours and days playing it. Although it was already possible to buy highly detailed, officially licensed plastic *Final Fantasy* figurines, Baldwin yearned for low-polygon figurines that more closely resembled the original low-detail graphics of early 3D computer games – a product that did not exist at the time. The artist extracted the 3D model data and textures from an archived copy of the game and painstakingly reconstructed and posed the models true to the original style and to his own personal

wishes. Shapeways is not only a print service provider, but also offers artists the option to sell their designs in an online store so that the company can print and ship orders on demand. Not unaware of the potential infringement of the intellectual property of Square Enix, the creator of *Final Fantasy*, Baldwin offered the figures for sale. He supposed Square Enix would not mind because the company had never sold the low-poly figures he and many other gaming fans were really excited about. He offered them in his Shapeways shop for a small premium on top of the raw production price with no expectation that this would provide compensation for the hours of free time he had invested in producing the models. There was never a financial incentive for creating his figurines; he simply wanted a product that no traditional manufacturer was offering on the market – arguably because the item was too niche to be mass-produced, even thought it would undoubtedly prove popular. In a short time, the models became an online success and Shapeways received a cease and desist order from Square Enix. They complied immediately, and as a result it became impossible for anyone to buy low-poly *Final Fantasy VII* figures – not even an officially licensed version. The legal aftermath left such a sour taste in his mouth that Baldwin did not wish to contribute any images of his beautiful models to this book.

Baldwin's creation is an example of fan art – the act of producing derivative artwork based on popular visual media out of sheer passion. Fan art often enters gray legal territory. In the case of Baldwin's figurines, the reaction of Square Enix is a classic example of how large corporations use copyright as a tool to protect their property, but in a way that also ends up stifling potential innovations that could lead to new sources of revenue and brand loyalty. In the words of RepRap founder Adrian Bowyer, "Suing one's customers is a crude and ineffective method of protecting one's business model."

The second tale is about Dutch kinetic artist Theo Jansen and demonstrates how a similar scenario can unfold without alienating the fan base. The artist has been creating his Strandbeests ("beach animals") since 1990, trying to invent a new form of life. These Strandbeests are enormous beings assembled from plastic conduit pipe. They can walk by themselves, harnessing the wind as a power source. Jansen always imagined that, in a distant future, his Strandbeests would go to the hardware store and steal or buy their own plastic tubing in order to reproduce. What he did not know was that, behind his back, his creations were already reproducing: fans were rebuilding them

are still designing new animals and 3D-printed accessories, such as a propeller that can be attached to the beast so that it can walk by harnessing the wind. In this case, the ability to copy is the ability to innovate. Bringing the Strandbeest project to a different level that also fits with the artistic philosophy of the original maker opens the creation to a much broader public and also generates shared revenue for all involved.

If the history of digital content creation and Internet consumption have taught us anything, it is that once the software for producing digital media becomes straightforward to use and accessible, derivative and remixed content quickly be-

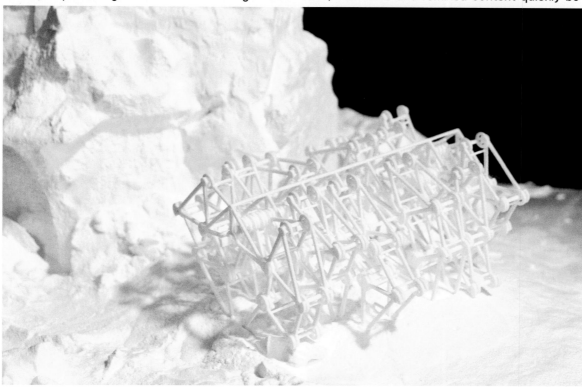

Theo Jansen, *3D-Printed Strandbeests* → **p. 168**

in various forms using DIY techniques. When two designers, Bo Jansen and Tim van Bentum, approached Theo Jansen with the idea of developing miniature 3D-printed offspring of his Strandbeests – models that would be able to walk straight out of the printer without any further assembly (a popular benefit of 3D printing) – the original creator was immediately exited. Together, the three now jointly develop a range of 3D-printed Strandbeests, which are sold in Theo Jansen's shop at Shapeways. The 3D-printed Strandbeest is one of the best-selling and most interesting objects available through the service. It is an impressive example of appropriation being embraced by the original creator. Bo Jansen, Tim van Bentum, and Theo Jansen

comes a popular form of creative expression that people like to share. This "remix culture" is one of the most iconic creative expressions to come out of digital media and the high potential for manipulation, appropriation, and sharing. There is nothing to indicate that this will be any different for 3D-printable online content. On popular communities for posting and sharing digital 3D blueprints, one can already see creators interacting with 3D files in a similar way to how content is constantly being remixed into new memes on YouTube.

Matthew Plummer-Fernandez, *sekuMoi Mecy* → **p. 90**

Information Wants to Be Free

Steward Brand, editor of the *Whole Earth Catalog*[A] (a user-generated catalog of counter-culture from the late 1960s that is now often referred to as the Internet in paperback form), proclaimed back in 1984 his famous maxim: "information wants to be free." It wants to be free because it has become so cheap (too cheap to measure) to copy, remix, and redistribute it. But on the other hand, Brand argues that "information wants to be expensive" because of its immense value in economic terms. It is this tension that forms the basis of the endless debates about value, authorship, and copyright that are now heating up again in relation to 3D printing. "Each round of new devices makes the tension worse, not better," Brand concludes.

A slew of tools have been invented over the years to protect the value of information so that it can be monetized by its creator. Instigated in the fifteenth and sixteenth centuries in reaction to the proliferation of the printing press and as a means to restrict the printing of books, copyright law protects the creators of intellectual wealth.

→A **Fall 1969 cover,** *Whole Earth Catalog*

Original writings, books, theater plays, music, paintings, photographs, and films as well as applied works such as the written source code of computer programs and the content of digital databases are thus protected against unauthorized copies being made. Copyright does not protect an idea or concept, only the specific expression of an idea. One does not need to apply for copyright protection; it is automatically applied to any original expression that is fixed in a tangible medium.

When the music industry was suddenly disrupted by Napster, which made the (mostly unauthorized) online file sharing of digital music a user-friendly experience, this was often a clear-cut case of copyright infringement because the files shared were carbon copies of the original music: they were exactly the same as the original. For 3D printing, the situation is very different because the vast amount of utilitarian three-dimensional objects are not protected by copyright. While functional objects can be protected by patents, these are not applied automatically and are far less likely to be encountered with tangible everyday items. It will also be much harder to stop the distribution of patent-infringing digital design since such infringement only happens when the item is physically produced, not when the blueprints are shared online. In the tale of the two 3D-printed fan art copies, the infringement is less clear and a court would have to settle the matter. Baldwin's figurines are possibly a clear case of copyright infringement because the creator extracted the 3D source code from the game itself, and that code is protected under copyright law in most countries.

Kiosk is a project which is intentionally wheeling into this gray area with the aim of fostering a discussion about authorship and ownership. The Kiosk cargo bike is fitted with a 3D scanner and a 3D printer and explores a near-future scenario in which these so-called digital fabricators are so ubiquitous that we see them on street corners. Real items can be digitized and printed just like downloaded digital files. In an act of street theater performed in the context of large design fairs and exhibitions, the project shows how the street will appropriate (sample, remix, improve, enlarge, shrink, or copy) objects in ways beyond the control of the original creators. When is the derivative a cheap fake, and when does it become a new piece? The Kiosk project also acts as a platform for ongoing exploration into the role of the designer when goods are moved around in the form of digital blueprints. Its ultimate aim is to challenge both authors and audience to come

up with new interpretations of intellectual property and authorship, and to create new scenarios around 3D printing in which both parties benefit.

New Opportunities

At the moment, 3D-printed copies do not pose much of a threat to traditionally manufactured goods because they have been designed for very different production processes such as injection molding and metal stamping. In order to make an item 3D-printable, it is necessary to redesign it completely with the respective production process in mind. The resulting copy is very different than the original product and often of lesser quality. It would be difficult to mistake the two. But could Pandora's box be opened if more and more products are designed for digital manufacturing and distribution in the first place? If someone is able to extract or reverse engineer the digital data, a 3D-printed copy will be as good as the original since both are produced in exactly the same way.

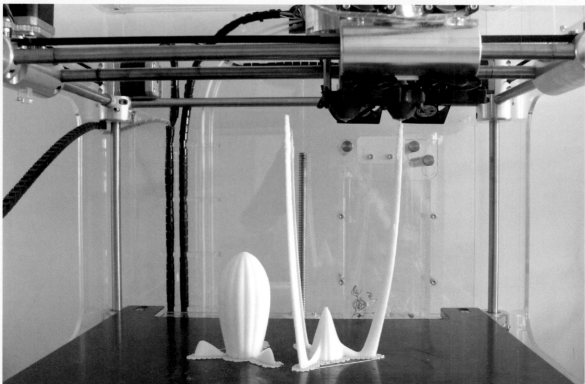

The concerns surrounding intellectual property with respect to 3D printing are justified, but it is important that those who are concerned do not stifle and banish the innovative and lucrative potential of the technology. It is vital that a constructive discussion about this issue be held before the technology becomes mainstream. Artists and designers can play an important role in visualizing and shaping new scenarios together with companies and other stakeholders. Reflecting on his own situation, Stuart Brand says, "If you cling blindly to the expensive part of the paradox, you miss all the action going on in the free part."

As regards 3D printing, it is not the machine itself that poses problems in terms of intellectual property, but the digital nature of the CAD files that are fed into it. When Brand talks about information that "wants to be free," he means the widespread access to data without obstacles, and this is where we have to look for innovative solutions. Since 3D printing applications behave more like software than like products, one can benefit from the numerous experiments with business models in the Internet industry, such as the freemium model where 90% of the product is given away (because data is cheap) and an increasing appification of products can be expected.

The product itself is not the most important factor, but the associated service provided. Some examples are already developing where, strictly speaking, there is no original – no singular iconic design piece that can be copied, but a family of forms. These are compiled one by one by an app, a service that creates a new personalized product every time someone interacts with it.

One such example is Kinematics by Boston design studio Nervous System. This is not a static object, but a framework that customers can play with to create their own jewelry, so every item designed is unique. It is not a universal design tool, and it is a far cry from traditional jack-of-all-trades CAD software. The software is open enough to let people play and create, but on the other hand it is specific enough to recognize the individual design language of Nervous System and the Kinematics product line. It is an "appified" product blueprint, and while one could certainly copy one of the product outcomes, there is little incentive to do so because there is no single iconic "original" that everybody wants. The added value for customers is created by the ability to play with the tool and customize personal items of jewelry. The online software is integrated with Shapeways to offer a complete customized product solution. To further reduce the incentive for product piracy, Nervous

System offers a second, simplified version tailored for at-home 3D printing. Users can export their designs for free to print on any available machine, but the options are more limited given the less wide-ranging technical possibilities of the baseline of consumer 3D printers. With this approach, copying the final product no longer makes much sense, because it would be necessary to copy the entire ecosystem.

Nervous System, *Kinematics* → **p. 250**

C

Interface and Interact

→ #3D scanning   → #customization   → #intangible
→ #interact   → #software

Unfold and Tim Knapen, *L'Artisan Électronique* → **p. 56**

Before you can print your own things, you need to design them. Of course you could always buy digital blueprints, but then you would miss out on half the fun. People often say that everyone will be a designer in the future given the grand vision of there being a 3D printer in every home. But how valid is this statement? Will everybody be a designer if and when 3D printing is an everyday part of life? It sounds simple: fire up your 3D modeler, draft your idea, then press print – just as you would with your trusted word processor. But unlike typing a text on a computer or even editing a photograph, drawing a three-dimensional object is not as straightforward. Even if the market is soon flooded with user-friendly design software for 3D printing, the question remains how many people will want to design and make their own items from scratch just because they can.

To illustrate this point, consider some of the machines and appliances marketed over the last 50 years that have promised to turn us into at-home producers. Both the sewing machine and the bread maker had the potential to shift power to the consumer by allowing people to make things themselves. With the help of some readily available patterns and recipes, common low-cost ingredients could be turned into customized home-made products with relative ease. But the sewing machine was mainly used to shorten the occasional pair of pants or to fix a hole in a shirt, while bread makers ended up gathering dust on kitchen shelves around the world. Joris Peels, former community manager at Shapeways, coined the term "the Singer problem" to describe this scenario. There are many different reasons why we still buy our bread and clothes ready-made, ranging from a lack of skills and free time to the power of marketing and our desire to have what everybody else has.

There is of course one major difference between 3D printing and the other two tools mentioned. Like other digital manufacturing tools, 3D printers function as a hybrid tool with digital and physical elements, the former adding automation and intelligence. When comparing a 3D printer with a standard sewing machine, then, the front end is different. Sewing your own clothes requires a form of manual labor as well as craft skills when it comes to designing a pattern, and cutting and

Bengler, *Terrafab* → p. 167

sewing the clothing. In order to physically produce something on a 3D printer, or to send a digital file to a 3D print service, no handicraft skills are needed; these only come into play when creating a digital design on the computer. And there's the rub: designing a 3D file for printing is a craft in its own right. Design skills have shifted from physical handicraft to virtual conception.

3D for Dummies

Virtual making could potentially pose a threat to 3D printing going into the future. Nevertheless, the digital front end of 3D printing, the design software, also offers many opportunities. 3D design software or modeling software has traditionally been very specialized and therefore out of reach for many casual designers both in terms of budget and learning curve. People who buy a cheap at-home 3D printer today often end up printing small items using digital files they have found online. The difficulty with these files is that they are not

very easy to modify or manipulate if you are not trained in using powerful 3D design software. In other words, it can be very difficult to customize a digital file of a doorknob, for example, to make it the perfect fit for your own doors at home.

Driven by the increasing popularity of 3D printing, a variety of free (or relatively cheap) and user-friendly applications have stepped in to fill the void. Software such as 3DTin, Tinkercad, and 123D Design offer various possibilities for less-skilled makers to start realizing their ideas on the screen and to create 3D files for printing.

Capturing Reality

Another way to create a digital 3D file is by using a 3D scanner to image a physical object and transform the data into a digital model. The conceptual idea of 3D scanning for reproducing a three-dimensional object – often models of people – dates back to the nineteenth century. In 1860 a Frenchman named François Willème filed a patent on a process called photosculpture. With the help of the new technique of photography, he invented a way to reproduce a three-dimensional copy of a person in a very short timeframe: The subject had to sit for ten seconds on a round platform that was divided into a series of wedges. While seated, 24 cameras, one for each wedge, took a picture of the subject from different angles. These photographic images were later projected onto a hump of clay, and Willème was able to create a person's sculpture by contouring the 24 projected photographs. In contrast to the conventional method, where models had to pose in front of a sculptor for many hours or days, the subject merely had to stand still for ten seconds – almost the same amount of time it takes today to carry out a digital 3D scan of a person.

3D photogrammetry is a 3D scanning technique used today which shares some similarities with the nineteenth-century photo sculpture process. Just like the latter, 3D photogrammetry uses a series of photographs, taken from various angles, to reconstruct a scene in three dimensions. 123D Catch is a free app that uses the same technique to turn a standard smartphone camera into a mobile 3D scanner, allowing users to scan people, objects, and places.

3D scanning is often seen as the holy grail for 3D printing. How easy would it be to scan an existing object, such as a doorknob, and then send it to the computer to print a replacement? Unfortunately, the scanning of objects has some flaws. While we humans have cognitive intelligence and are able to figure out the intended purpose of unfamiliar objects and to distinguish between what is important and what is not in terms of function, a computer lacks these skills and can only copy the surface of an object as faithfully as possible. With most accessible 3D scanning solutions, a fair amount of post-processing is necessary to make the model almost as good as the original. The Replicants project by Lorna Barnshaw aims to show these flaws and highlight digital artifacts in the output of simple 3D scanning methods. She imaged her face using several 3D scan solutions, including a VIUscan 3D scanner, 123D Catch, and Cubify Capture and then printed the data without any post-processing clean-up. The results are distorted mask-like objects that show the transition from material to virtual and back to material, presenting the computer's interpretation of the subject as seen by digital eyes.

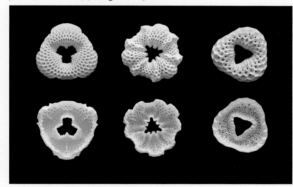

In-Flexions, *KiLight*   → **p. 148**

Fluid Things

With more people gaining access to digital production tools such as 3D printers, an interesting approach for designers is to start focusing more on the design process rather than on the final physical end result. The flexibility of digital data offers various ways to interact with it, creating new opportunities. Furthermore, products exist only as digital data for much of the life cycle because they do not materialize until the final moment, in many cases, when money changes hands. There is a long period in which data stays malleable or fluid.

Software, therefore, is not merely a means to the end of designing a final physical object, but an essential part of the design itself.

One approach to interacting with digital data is to create designs that can be customized by end users. Systems that allow people to customize a specific product are not new. It is a marketing technique that has often been experimented with in the creation of consumer goods. The philosophy behind customization is that making items unique increases their value. Like in the world of art, products are valued more if only one or a few of them are made. On the other hand, customization responds to the consumer desire for self-made objects. Things that people make by their own hand or that have been crafted by a dear friend are not thrown away so readily. Traditional products do not lend themselves well to customization, and there is a very real penalty for every extra choice. Most customization options in traditional manufacturing work by offering customers a variety of colors, materials, and sometimes shape, and the chosen elements can then be combined into something new. However, each option then has to be mass produced, distributed, and stocked. This leads to exponentially rising costs. As a result, customization is limited to selected products, often with only a few options. In contrast, since a product for 3D printing is digital until it is made material, it can be customized at will without any penalty. Every product is printed on demand anyway.

Online 3D print service providers perceive customization as a major opportunity to reach a larger public. Most of these services offer a couple of customization tools on their web site and showcase various designers and software developers who have produced their product in the form of customizable fluid designs. For example, a simple app that allows users to customize a replacement knob for an appliance by moving sliders to adjust the size before selecting various options. Or the classic example of adding a name tag to a product. However, instead of letting the customer tweak one or more parameters at will, designers who embrace the idea of fluid design are constantly searching for more sensible and poetic parameters as well as more natural ways to interact with them.

All kinds of tangible and intangible data can be used as parameters when customizing objects. Everything that is digitally recorded, registered, or created can also be used as input when generating a digital object, be it someone's favorite music, the place they live, or their Facebook activity. This data is malleable; it can be interpreted, translated, tweaked, and manipulated into different shapes and objects. Fung Kwok Pan's Fluid Vase, for example, allows customers to model the motion of liquid being poured into a vase. A single moment of this action can then be frozen in time and printed as a vase. Terrafab from Bengler allows users to select a square section of Norway and download the altitude map of the area as a 3D model or send it straight to an online print service. Loci from Andrew Spitz is an app that interprets more personal data to create objects: users enter their air travel details over a certain period to create a personal flight sculpture. Conveniently, the app can also pull data directly from popular flight tracking services. The resulting flight sculptures can be kept as a souvenir of past travels or used as a small personal infographic.

Front, *Sketch Furniture* → **p. 130**

Physical Interfaces

In all of these examples, the computer screen and mouse are used to interact with the digital object. Part of what makes 3D design software more difficult to master than, for example, a word processor or image manipulator is that it has to emulate a process that is normally tactile and spatial is that it has to emulate a process that is normally tactile and spatial – like modeling clay – in a two-dimensional space (the computer screen).

Whether we are checking our e-mails, drafting an architectural drawing, or designing a doorknob, the graphical user interface and the way in which we interact with the screen are no different. There is a very limited relation between what the hand does and the final product: we do not directly manipulate digital model. In general, we type, scroll, and drag, but we never use tools such as a hammer, knife, or ruler to design a three-dimensional object on the computer. When we use a computer, we give commands instead of carrying them out because the computer does the work for us. The illogical relation between the graphic interface of a computer and its physical operation has often been pointed out. A couple of projects in this book explore an interface that tries to redefine the physical actions used when processing digital data with the aim of adding some sense of space and touch in the digital realm. To quote Malcolm McCullough, author of *Abstracting Craft*: "Little can surpass the hands in showing that we know more than we can say."

One undertaking that literally transforms a traditional hand tool into a digital tool is Kathryn Hinton's Digital Hammer project. Trained as a metalsmith, she explores the merging of traditional and digital tools in the craft of metalwork. She created a digital interface that takes information from the physical strikes of a haptic hammering device and feeds this directly into a 3D computer design program. The resulting digital design, which shows a clearly digital aesthetic, is then materialized again using digital manufacturing techniques such as 3D printing and CNC milling.

Another very hands-on device was used by the Front design team to conceive their Sketch Furniture. Front wanted to directly materialize freehand 3D sketches, retaining their spontaneity. To do so, they used 3D motion capture hardware from the movie and game industry to track the freehand movements they made in midair with a special pen. These movements or spatial sketches were later turned into a 3D model by giving the paths a three-dimensional brushstroke as if a material was oozing out of the pen. Finally, the objects were printed, turning the one-stroke sketch directly into a material object.

Both projects look at the digital interface from a new angle, trying to link the action of the body with the generated data. However, they do not introduce the audience or a crowd within the design process of an object. Other projects try to open these more physical interfaces to a wider public. Design studio In-Flexions, for example, produced Ki-Light, which uses the movements and colors of someone's body, hands, and arms to generate 3D-printable lampshades. A similar approach can be found in L'Artisan Électronique, an installation from Belgian design studio Unfold in collaboration with interaction designer Tim Knapen. L'Artisan Électronique is a modern-day take on the traditional artisan pottery studio and involves both the design process (a virtual potter's wheel) and production (a ceramic 3D printer pioneered by Unfold in 2009). The digital potter's wheel lets visitors 'turn' a pot in thin air. By moving their hands in front of a rotating virtual model, they can 'mold' the digital material as if they were handling virtual clay. The digital pot can then be 3D-printed in clay using the ceramic 3D printer that is part of the studio setup. The installation visualizes the whole creation process, from the emergence of the design to its actual realization.

Both projects are conceived as installations for public spaces, mostly museums, encouraging visitors to experiment with the tools and to create and refine their own objects. It would be easy to convert this technology into customization tools with every lamp or vase then being produced for the person who designed it. In any case, these two projects take a slightly different approach — one of co-creation in which the designer of the installation and the people who interact with it in a museum work together to design a family of forms. The designers establish the framework but crowdsource the final decision to the public. In the end, the designs are made by various people, but they all undeniably belong to the same family with a formal language instigated by the designers.

Nevertheless, this profession will not disappear; it will most likely shift from someone who designs finished objects to someone who designs processes and systems. We will see a gradient of different sorts of designers emerge, from hobbyists and prosumers to semi-professionals and established experts — all of whom will have the resources and capability to manufacture, which used to be the reserve of those happy few who landed a job in the industry. It is not the case that studying languages and mathematics in school has turned us all into brilliant novelists and engineers, but we can all understand and appreciate the subjects in broad terms. In a similar way, a greater focus on design in education will lead to a better appreciation of the factors involved and to new situations where designers and consumers can work together to create products using fluid design software services.

Everyone a Designer?

The statement "In the future, everyone will be a designer" can be traced back to Chris Bangle, the former director of design at BMW, who argued in 2005 that in a future with ubiquitous access to 3D printing, design would become just as important as literacy and would have to be taught in schools. Most high schools used to teach craft and design, but with the shift to the post-industrial knowledge economy over the last couple of decades, this subject was scrapped from most curricula. Bangle argued that when everybody was a designer, this would put pressure on the design profession itself.

Lorna Barnshaw, *Replicants* → p. 208

D

Body Topology

→ #3D scanning   → #body topology   → #medical   →#wearables

The Greeks were very suspicious about looking at their own reflection. It was believed to bring bad luck, or even to be fatal. Skip forward a few centuries and you will find that nobody shares this sort of superstition anymore. With a digital camera in almost every cell phone, laptop, and tablet, it has become common to see your own image appear everywhere. In 2013 the Oxford Dictionaries Word of the Year was "selfie" – a clear demonstration of the fact that we are not very superstitious about our own image anymore. On the contrary, this word (used to describe a casual self-portrait taken with a digital camera or smartphone camera, usually in a mirror or at arm's length) has become increasingly popular. Selfies can be found all over the Internet, and even the Obamas and the Pope have been caught taking such pictures. In the same way as Narcissus fell in love with his own reflection, those in the digital generation seem to be in love with their own image.

The Out-of-Body
Experience

Ten years ago body scanners were still the domain of research institutes, but today 3D scanning is becoming ubiquitous. Shortly after its release, Microsoft's XBox Kinect motion sensor – a device normally used to detect a gamer's body movements – was hacked to access the raw data from its 3D camera sensors. Several 3D scanners have since been developed using a hacked Kinect. Impressed by the creativity and enthusiasm of the hackers, Microsoft assumed the official position that it would support the use of the Kinect for such creative purposes.

123D Catch is a smartphone app that does not even require special additional hardware. It generates 3D data by analyzing and combining multiple photographs of the same object. On popular 3D file repositories like Thingiverse and Shapeways, there are already numerous examples of people who have scanned parts of their body or their face and made the data available to the wider community to print. Similarly, a crowd-catcher at many of today's 3D printing events and trade fairs is a 3D scan booth that allows visitors to digitize themselves and print out a mini-me.

There is nothing strange about people being mesmerized by this technology. For the first time in human history, we are now able to see ourselves as others normally see us, full color and in three dimensions. We can take a picture of ourselves or stare in the mirror for hours on end, but never before have we had the chance to see ourselves completely, from all angles. It is as if we could step outside of our bodies and see ourselves from a new perspective. This is a brand-new experience; making a miniature model – a 3D "selfie" – gives

blablabLAB, *Be Your Own Souvenir* → **p. 70**

a new meaning to how we see our own body. We can only begin to wonder what the word of the year will be in 2023.

Be Your Own Souvenir from Barcelona-based studio BlablabLAB was one of the first projects that incorporated the idea of a 3D photo booth. BlablabLAB developed an open-source 3D scanner based on the Kinect and connected this to a MakerBot 3D printer so that the scan could be 3D-printed on the spot. The most interesting aspect of this project is that it invites people to pose in a public area, such as a museum space or on the street, when the scan is made. The process takes a couple of minutes to complete, and during this time the volunteers perform as a living sculpture, standing still on a platform so as not to ruin the scan by moving. Fifteen minutes later, the result of this short performance is printed in brightly colored plastic as a souvenir of their time.

BlablabLAB does not focus as much on the precision of the scan or the print as on the performative act of the 3D scan and the magical moment of an instant snapshot becoming tangible: the 3D printing process is akin to the slow development of an instant Polaroid photograph.

In contrast, Omote 3D in Japan focuses on the photorealistic accuracy of the replica. Miniature prints of humans by Omote 3D are highly detailed and printed in color, making it much easier – and almost surreal – to recognize someone. If Be Your Own Souvenir is the Polaroid version of a 3D image, this second project can be considered the professional 3D photo studio.

While a 3D copy of your own body might resemble an out-of-body experience, 3D printing can also help to materialize the most precious in-body experience, that of an unborn child developing in the womb. For his 2009 Fetus Project, Jorge Lopes dos Santos used ultrasound scans to capture data from a mother's womb in order to recreate a three-dimensional representation of the fetus. In this way, soon-to-be-parents can now hold their unborn child.

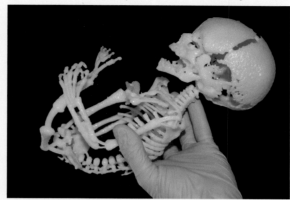

Jorge Lopes dos Santos, *Fetus 3D Project* → **p. 202**

Although this project could be helpful in some specific medical and educational contexts, such as allowing visually impaired parents to experience the development of their baby, it also raises many ethical questions. Sculptures of unborn babies are powerful: for some they are heart breaking, for others disturbing. It is only a recent phenomenon that we can see our unborn child at all, and this technology is still evolving from the grainy black and white sonograms of 30 years ago to today's popular 3D imagery. How far will technology go beyond medical diagnosis as regards the imaging of an unborn child? Does it become uncanny and take away the magic from that first moment of holding a newborn baby? Can the materialization of an unborn bring any additional happiness, joy, or comfort to our lives?

Medical Applications

There is no canvas more diverse than the human body, and every functional or aesthetic object that needs to interact with the body is better suited when customized for an individual body. Uniqueness and customization is something that 3D printing excels at. The medical world in particular is an important field where 3D printing could really make a difference. Researchers are currently investigating how the technology might be applied for printing organs and bodily tissues. If they successfully print and transplant an organ into a human body, many lives could be saved, but many more ethical questions will arise. Are we playing

PARTYLab, *Omote 3D Shashin Kan* → **p. 68**

God by copying our organs? What else would we attempt to clone in the future – whole humans, maybe?

And while researchers are working in labs on machines that can print bodily tissues like organs, 3D printing is already being used for prostheses and implants. Tailor-made implants for jaws and prostheses for the face could be modeled onto the patient's body with the same precision as the original lost parts. Other prostheses could be made more cheaply and yet with personalized details, such as 3D-printed glass eyes.

One project that shows the impact of this technique on a human life is the story of four-year-old Emma, who was born with arthrogryposis multiplex congenita (AMC), a disease that affects the joints and muscles. Emma's arms are so weak that she is not able to lift them by herself. A robotic exoskeleton could help her to move her arms much better. Unfortunately, these exoskeletons are made from metal and are too heavy for a young girl to use. The hospital that was treating her decided to create a 3D-printed anchor vest for the girl. The plastic parts of the skeleton, which Emma calls "magic arms," are lightweight and enable her to use both her arms. If a part of the suit breaks, the hospital can print it again, and while Emma is growing it is easy to print a slightly larger version time after time. In this field, 3D printing really makes a difference because every device needs to be customized for the patient.

Some 3D-printed medical appliances are not only functional, but also have aesthetic value. Bespoke Innovation designed the Fairings, 3D-printed aesthetic covers for existing prostheses that can be personalized depending on the preferences of the wearer. There is also the Cortex 3D cast, an alternative for the splint or plaster cast for broken bones. The Cortex 3D is derived from a 3D scan of the broken arm or leg and is then 3D-printed, making a ventilated, comfortable, light and thin cast that also looks much better than plaster. These two medical applications do not want to be hidden from sight like their traditional counterparts; they want to be seen.

3d Systems and Bespoke, *Prostheses*

Wearables

It is a small step from practical but visually impressive medical applications to the world of fashion, accessories, and jewelry. In the same way that medical objects can be designed with a perfect fit for an individual's body, jewelry and fashion can be, too.

Because of its size limitations and a fairly high price per volume, 3D printing lends itself extremely well to the production of small luxury goods like jewelry, for which people are more willing to pay premium prices. Jewelry also tends to be very personal, and the shapes that can be made are endless. Designers like Maiko Gubler, Nervous System, and Dorry Hsu are exploring what it means to be a goldsmith in the digital age by forging digital matter.

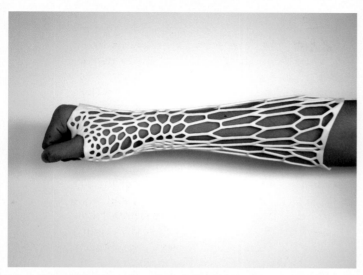

Jake Evill, *Cortex3D*

A good example in the world of fashion is Dita's Gown, designed by fashion designer Michael Schmidt and architect Francis Bitonti. Similar to how the Cortex 3D is designed based on a 3D body scan, Schmidt and Bitonti used a digital model of burlesque dancer Dita von Teese to create a dress that is perfectly shaped to match her body. The concept of the gown is more like armor than a traditional dress, as most parts are stiff. This rigidity emphasizes the form, most notably the dramatic sculptural shoulder pieces. Other parts seamlessly change into a more flexible chain mail-like mesh with openings of a varying size so that the dress can move with the body.

At first sight, 3D printing does not seem very appropriate for fashion items. When we think of 3D printing, soft and flexible materials do not come to mind. That explains why Dita's Gown resembles chain mail armor. Chain mail was made from a lot of small, stiff parts, such as iron rings, put together in a way that allowed each part to move separately. In the medieval days, such an intricate design was the result of painstaking work by hand, but 3D printing means that the same type of garment can be produced in one go, without the need for assembly, which saves considerable time.

While Dita's Gown was produced using the chain mail method in order to add flexibility to an otherwise rigid material, other experimental items have been made from new 3D printing materials that are inherently flexible. For her 2013 haute couture collection Voltage, Iris van Herpen worked together with Neri and Keren Oxman from MIT Media Lab and Austrian architect Julia Koerner to create an outfit with both rigid and flexible sections of material. Neri Oxman's Mediated Matter research group at the MIT Media Lab is investigat-

ing materials with different features and specifications that can be used for 3D printing. They look at principles in biology and try to create similar results for printed matter. In the outfit they created together with Iris van Herpen, the material changes seamlessly between rigid and flexible where needed. In this sense, the material characteristics stipulate not only the form, but also the way in which the fabric should move.

The same principle is used for Marloes ten Bhömer's 2010 Rapidprototypedshoe. This shoe comprises two materials with different microscopic structures, making it possible for the same shoe to be both rigid and flexible. Used in combination with meticulous measurements and scanning tools to image an individual's feet and manner of walking, 3D printing can offer advanced, sophisticated footwear for all kinds of people. Possible applications range from sports shoes to orthopedic shoes.

Marloes ten Bhömer, *Rapidprototypedshoe*
→ p. 228

With this in reach, we will probably see the one-size-fits-all philosophy being replaced by a highly personalized approach that allows people to buy custom-made shoes and clothes that perfectly fit and support the body. One could argue here that customized shoes and clothes already exist, and that is obviously true – tailored items have always existed. But with 3D scanning now becoming as straightforward as posing in a photo booth, customization is something that will soon be within reach for the masses, no longer the reserve of the privileged.

Francis Bitonti, *Dita's Gown* → **p. 248**

E

The Aesthetics of Complexity

→ #assembly free   → #complexity

Judging by the wide coverage it receives in the media today, one could be forgiven for thinking that 3D printing, more formally known as additive manufacturing, is a very recent invention. While the colloquial term "3D printing" has certainly increased in usage since 2011, the technology of additive manufacturing was born in the mid 1980s and was better known under the moniker "rapid prototyping" until very recently. In those days the structural quality of 3D-printed parts was unsuitable for creating functional parts, but it was good enough for visual prototypes. It gave manufacturers the chance to physically examine and even test a product before committing to mass manufacturing using conventional techniques. The rapid speed at which one could go from a sketch to a physical mock-up and the extreme versatility of 3D printing quickly made it the go-to choice for producing industrial prototypes. Because of its use as a prototyping tool, 3D printing had to imitate existing mass manufacturing techniques such as injection molding and machining, which also meant that it was bound by the technical and formal limitations of the same methods.

From Rapid Prototyping to Rapid Manufacturing

In 1990, Materialise was founded as the first 3D printing company in the Benelux region. It focuses on the research and development of specialized 3D-printed products and services for the medical and automotive industries, among others. Materialise soon realized that the technology had the potential to grow beyond serving as a method of creating industrial prototypes and niche technical items, and that it could one day be used to manufacture consumer products. The company thus founded .MGX by Materialise in 2003, a pioneering brand focused on the design and production of consumer products created using 3D printing methods. They invited world-class designers to experiment with 3D printing and come up with novel products that were only possible with this new technology. Released from its duty to imitate existing manufacturing methods and helped by eager designers, 3D printing could finally unlock its own formal language and aesthetic identity.

.MGX by Materialise was not the first to propose 3D printing as a manufacturing technique in its own right: designer Ron Arad had already

Janne Kyttanen, *Lily.MGX*

Patrick Jouin, *One_Shot.MGX* → **p. 111**

when an object is made more complex. On the contrary, in some cases there may even be a benefit. If a thousand holes are punched in an object, for instance, 3D printing saves material and production time compared to a conventional process. With 3D printing, designers and artists can explore new kinds of highly complex and intricate forms that would have been impossible to realize with traditional techniques, and these come at no extra cost. It is a proverbial candy store of new formal possibilities, resulting in a new design language that is baroque and often eclectic.

Single Shot: Assembly-Free Construction

To overcome the limitations of traditional manufacturing, traditional products are almost always assembled from simpler sub-elements that are easier to produce. With 3D printing, there is no inherent difference between printing all the elements of a wind-up toy car separately or already assembled in place, ready to go after the print job is done. This is one of the more practical implications of the paradigm "complexity is free."

One of the designers who played an instrumental role in the early history of .MGX by Materialise was Janne Kyttanen, who had just founded Freedom Of Creation, a design studio exclusively dedicated to the exploration of what is possible with 3D printing. Some of Kyttanen's first experiments were 3D-printed fabrics and chain mail constructed out of small, interlocking rings printed in one go. Traditionally, metal chain mail was labor-intensive to produce, entailing the manufacture and assembly of thousands of open rings that were then connected one by one and welded shut. 3D-printed chain mail mostly serves as a powerful demonstration of the assembly-free capabilities of 3D printing (the consumer interest for chain mail is niche at best), but other more refined assembly-free designs have emerged, such as the foldable One Shot stool created by Patrick Jouin for .MGX in 2004. In one simple action, this can be transformed into a compact tubular object for easy storage. The stool has a circular joint and a hinge. When the hinge is lifted, this joint turns and collapses the legs with a simple twist. As its name suggests, the One Shot stool is produced in one go, ready for use. Other examples of functional

presented his *Not Made by Hand, Not Made in China* collection of 3D-printed products in 2000. While Arad's collection was more of a concept or vision, .MGX by Materialise was one of the first companies to bring 3D-printed consumer goods to the market. The brand regularly releases new collections of 3D-printed design products, most of which are lamps and limited-edition pieces of furniture such as chairs and side tables. Looking at the very first collections, which are now already labelled "classics" by .MGX, one can already trace the major trends in the search for aesthetics in 3D printing.

Complexity Is Free

In his book *Fabricated: The New World of 3D Printing*, Cornell University researcher Hod Lipson describes ten of the underlying principles fundamental to 3D printing. The first principle he notes is that "manufacturing complexity is free." Unlike traditional manufacturing processes, where extra complexity requires a more expensive mold with more parts, there is no penalty with 3D printing

Ilona Huvenaars, *Knitted Vase* → p. 118

aesthetics are the Knitted Vase by Ilona Huvenaars and Willem Derks, which features a "knitted" neck that can expand when a bouquet of flowers is placed into it, and Jointed Jewels by Alissia Melka-Teichroew, in which several interconnected ball-and-socket joints are 3D-printed in one piece, creating a collection of aesthetic jewels.

Biomimicry

Nothing is more geometrically complex than nature, and many of the early 3D-printed consumer product experiments were often inspired by the environment. Janne Kyttanen's Lily from 2002 highlights the beauty of a water lily. In the same series he also designed the Lotus, the Dahlia and the Palm. Patrick Jouin's Bloom lamp mimics the way a flower unfolds during the day and closes again at night. Like the One Shot stool, the lampshade is printed with its mechanisms already in place.

While natural structures often have a very complex geometry, this complexity is often governed by simple mathematical rules. This duality has influenced many designers who, inspired by nature, program parametric objects that rec-

reate complex organic structures. By describing the object in code, one can almost literally breed very complex shapes, and a plentitude of variations can be generated by tweaking the parameters. This more open, abstract interpretation can be found in the Fractal table from Wertel Oberfell. This piece of furniture is the result of studies into fractal growth patterns in nature, and fractal algorithms were used to generate the form. Similar to the growth of a tree, stems grow into ever smaller branches until they become a dense surface that forms the tabletop. The work of Nervous System shows a similar generative approach, with algorithms drawn from nature shaping physical products. Their generative processes are sometimes embedded in user applications that allow consumers to interact with them as if they were playing in a virtual natural science lab.

Biomimicry is a field that looks at the models, systems, and elements of nature for the purpose of solving human problems. The goal of the Mediated Matter research group at MIT Media Lab, headed by architect Neri Oxman, is to investigate

Wertel Oberfell, *Fractal.MGX* → p. 122

the lower-level processes by which nature constructs matter itself, for instance how silkworms build their cocoons. The goal of Mediated Matter is to develop design principles inspired by nature and implement them in the invention of new 3D design and printing technologies.

Spatial Patterns and Mathematical Structures

Many popular 3D-printed artifacts today exploit the freedom of complexity and show an abundance of ornamentation. These ornaments can often be traced back to historic patterns found on fabrics, lace, or geometrical tiles. With the use of design software, these patterns can be digitally superimposed over other three-dimensional shapes, ranging from figurative forms to complex mathematical models. A striking example of this is Crania Anatomica Filigre, a piece by Josh Harker that was also a very successful Kickstarter project and a best seller on Shapeways. It is a three-dimensional incarnation of the colorful patterned ornamental skulls used in the Mexican celebration of Día de Muertos, the Day of the Dead. A human skull is overlaid with a decorative tattoo-like pattern which is then cut away, leaving an object with a very open structure that resembles lace, woodcutting, or filigree, and clearly refers to these old, labor-intensive crafts.

For some reason, it is in our nature to value complexity as precious. This is probably because for millennia it cost considerably more time and effort to produce complexity. It is hardwired in our minds to associate visual complexity with the dexterity and perseverance of the skilled craftsman. By using 3D printing to achieve these effects, the maker can trick the observer, playing on the intrinsic notion of value. On the other hand, people are not yet trained to value the labor that goes into coding a piece of software, for example, or crafting an object with digital technology. Over time, our ability to recognize and value these aspects of digital design will hopefully grow, and most people will be able to distinguish complexity for the sake of complexity from genuine digital craftsmanship.

Josh Harker, *Crania Anatomica* → **p. 115**

More Is a Bore?

Because complexity is free, it will continue to be a popular source of inspiration for creators. Just like everything else that is free, however, there is also an inevitable risk of overuse: "Just because you can, doesn't mean you have to." Will there be a time when ornamental complexity, traditionally associated with the skilled hand and patience of a craftsman, will shed this association and be seen as nothing more than excessive showing off?

Ultimately we are still in the early stages of the search for aesthetics in 3D printing. Many of the experiments we see today may appear outdated in ten years, but they are playing an important role in paving the way. With an increasing number of designers, artists, and makers gaining access to 3D printing, a mature formal language will develop over time, uniting and exploiting the full potential of the technology's aesthetic powers.

F

Building Blocks

→ #architecture   → #building blocks

"I'm trying to connect this, um, into here, and then make it like a car, and connect this into there." This answer probably sounds familiar to any parent who has seen one of their children fumbling with construction toys and asked what they were trying to do. In this case, it was a four-year-old boy who had grown frustrated trying to build a car out of the different brands of universal construction toys he owns, and which had most likely all ended up mixed together in one large tin. The scene is taken from a video that accompanies the Free Universal Construction Kit, a project developed by F.A.T. Lab and Sy-Lab. It offers the free download of nearly eighty 3D-printable adapter bricks that enable interoperability between a range of the most popular construction toy brands, including Lego, K'Nex, and Tinkertoy. Each brick features a male connector of one brand on the top side and a female connector of another brand on the bottom side. The kit allows children to play more creatively, unhampered by proprietary vendor lock-ins.

The Free Universal Construction Kit is first and foremost an artistic statement on intellectual property rights and grassroots fixing. However, it is also a great visual example of the capacity of 3D printing for producing custom-made adapters and connectors between a plentitude of already existing objects and components. Items are brought together that would not normally interact in any way unless duct tape was involved. The result may be highly individual one-off solutions where the same need will never arise again. In other cases, like the Free Universal Construction Kit, there may be commercial, economical, or other reasons why such parts do not exist.

A Shared Language for Things

An ever growing number of people will discover the world of digital fabrication and start creating adapters between incompatible things, replacement parts, improved components, and even full-fledged products. What is missing in the growing DIY community is a universal standards system that everyone can adhere to in order to facilitate the sharing of components, and to foster online collaboration between groups of designers and makers worldwide. This would create an open and dynamic layman's alternative to the proprietary industry standards issued by the International Organization for Standardization (ISO).

The aim of the OpenStructures project initiated by Belgian designer Thomas Lommée is to develop such a universal and open standard for things. By adhering to the standard's modular specifications in the design and manufacturing

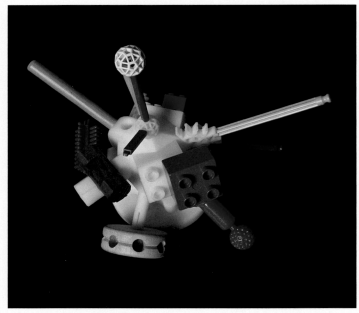

F.A.T. Lab and Sy-Lab, *Free Universal Construction Kit* **→ p. 169**

Thomas Lommée /Matt Sindall, *Brick MS*

Jesse Howard, *Transparent Tools*, grinder → **p. 102**

Hybrid Manufacturing

The first generation of 3D printing technology in the 1980s was only suitable for the creation of visual prototypes, but today's generation of machines are very capable of producing structural and attractive parts for consumer products in a variety of materials, including durable metals. This does not mean that people will be printing large-scale items like furniture in large quantities just yet, however. The price of 3D-printed items is not very competitive with respect to traditional manufactured items. On top of that, there is the bounding box dilemma: every 3D printer has a limited build volume, defined by the bounding box in which the part has to fit.

One approach to breaking out of the bounding box is by not relying on 3D printing for the manufacture of the entire object, but just the more complex parts where it makes a difference, using other production methods for the rest. Such a hybrid manufacturing approach combines the advantages of both 3D printing technology (highly customized parts) and the mass manufacturing industry (cheap standardized parts).

Swedish/German designers Clemens Weisshaar and Reed Kram used this strategy and created customized software to analyze the stresses in their Multithread line of furniture in order to modify and reconfigure the joints algorithmically. Once a design is finalized, the software generates the digital blueprints. These consist of a set of joints, printed with metal laser sintering, onto which computer-cut lengths of metal tubing are welded. The final pieces are polished and lacquered, creating a seamless transition between 3D-printed and standard parts. Each joint is customized with a bright computer graphic – a gradient that visualizes the otherwise unseen forces as they act within the joints.

Instead of seamlessly integrating 3D-printed and conventional parts like Kram and Weisshaar, the contrast between 3D-printed parts and wooden beams becomes an essential aspect of the aesthetic in the Keystones furniture by Minale Maeda. The work of Minale Maeda researches the potential of "multi-directional material translations" – hybrids of digital and analogue materials and techniques used in combination with open-source schematics and new distribution models. The Keystones project takes out the hard part of woodwork, the creation of professional joints, by using plastic joints. These are printed at a local manu-

process, designers can create products, parts, or components which are intrinsically more interoperable and therefore part of a larger ecosystem of things. One such designer is Jesse Howard, who created Transparent Tools, a set of domestic appliances that users can produce, repair, and modify by themselves. The range includes a toaster, coffee grinder, vacuum cleaner, and electric kettle but you will find none of these on the shelves of your favorite retailer. Instead, they come in the form of an instruction leaflet which features an exploded view and a list of readily available hardware products that can be either bought online (eBay links are provided) or salvaged from discarded appliances. To assemble all the self-sourced components, Howard designed a set of 3D-printable parts using the OpenStructures grid and made them available online. By sharing these files with a Creative Commons license, the designer invites others to appropriate and evolve his designs or to repurpose the parts in entirely different products.

facturing center, and standard pieces of wood can be fixed into them by the user. Furthermore, there is no need for an assembly manual because all the instructions are embossed on the surface of the joints themselves. Each piece features just a single connector (hence the name Keystone, referring to the central stone at the apex of an arch), and it includes the fasteners – large plastic bolts 3D-printed inside their socket and ready to be screwed tight. The Keystone joints become the essential part of the furniture, while customers can make the rest by hand.

Breaking Ground Outside the Bounding Box

Designing hybrid manufactured products that combine the strengths of traditional and digital manufacturing techniques is one way to overcome

the limitations of the bounding box. The other is to use brute force and build a bigger printer.

The struggle with the scale limitations of 3D printing technology is at the epicenter of the race to build the world's first habitable architectural structure built entirely using 3D printing. The race informally began in early 2013 when a number of contestants unveiled plans in short succession to build the first 3D-printed house. All the teams share the same goal but propose very different methods to achieve it, and some teams even publicly dispute the claims made by others. On thing is certain: none of the proposed techniques have been tried and tested before on this scale. To quote DUS Architects, one of the contestants: "We are laymen, and we are learning by doing."

However, the dream of 3D-printed buildings is not new. A couple of pioneers have been steadily working on the necessary technology to print buildings since the end of last century. Since 1998 Dr. Behrokh Khoshnevis of the University of Southern California has been developing a layered manufacturing process called Contour Crafting, in which cement or concrete is pumped through a nozzle connected to a computer-controlled crane or gantry. This draws the contours of the large-scale structure to be built layer by layer. Whereas the Contour Crafting project is mostly developing sophisticated extrusion systems and focusing on small-scale proof-of-concept setups, Enrico Dini, a passionate Italian inventor and self-proclaimed "stone alchemist" has made it his life-long project to build the world's largest 3D printer. In 2008 Dini pioneered a method for printing sandstone using a process that was inspired by the formation of rock in seawater. His machine is a giant six-by-six-meter gantry that sprays a liquid binder over layers of sand mixed with magnesium oxide, in a similar way to how an inkjet printer sprays ink over a sheet of paper. Dini can stake a claim for having produced the first-ever printed architectural structure, an archetypically shaped one-room house designed by Marco Ferreri in 2010.

Today Enrico Dini is aiming for even greater things than the first 3D-printed house on Earth. He has teamed up with the European Space Agency and renowned architects Foster + Partners to test the feasibility of 3D-printed permanent moon bases built out of moondust. At the same time, Contour Crafting is also aiming for the moon in a partnership with NASA.

Given the significant challenges of scaling up 3D printing machinery to encompass an entire building, many including Dini have concluded that, for the time being, the most pragmatic approach

Foster + Partners, *Habitable Lunar Settlements* → S. 192

thetic co-development of Canal House and what is known as a "construction site 2.0."

The 3D Print Canal House is being printed on-site with the KamerMaker, a shipping container that has been converted into a giant 3D printer and that shares some of its DNA with the Ultimaker desktop 3D printer. The building will be printed in sections up to 3.5 meters high covering an area of up to two square meters. These sections are first assembled into rooms and then stacked one by one to form the final house. Each room will be different with its own unique characteristics, ornamentation, and smart features such as shading scripted to the exact angles of the sun in that location. Cavities in the hollow walls provide a space for all the usual conduits and pipes.

is to fabricate constructions in sections and then to stack these sections on-site. One could argue that traditional masonry, the act of stacking bricks on top of each other to build a house, is essentially an additive manufacturing process in itself.

Brian Peters's Building Bytes relies on the age-old art of laying bricks, but he imagines the bricks themselves to be dynamically generated by software and printed in clay on a farm of standard 3D printers. Each brick would be uniquely shaped and placed in a specific position as part of a complex architectural structure. Peters is also part of the team at DUS architects, the studio led by architects Hans Vermeulen, Martine de Wit, and Hedwig Heinsman. This team is well on its way to making the claim for the first 3D-printed house using a large-scale mobile 3D printer, the KamerMaker ("RoomBuilder"), which was developed in-house.

In January 2014, DUS architects met with their international partners to break ground on the construction site of the 3D Print Canal House in Amsterdam. Located on one of the city's most famous canals, this site will most likely be home to the first 3D-printed house. It will also act as an exhibition and interactive research center for 3D-printed architecture and related areas, such as material recycling, policy-making, and smart electricity grids. The aim is to engage the community and other stakeholders in the conceptual and aes-

Smart Matter

Instead of depending on large 3D printers to construct buildings with prefabricated sections or in one piece, we might in the future rely on swarms of small robots that construct buildings in a similar way to how ants build their colony. An even more utopian scenario is already in the making at various research labs: programmable matter is the name for tiny units of matter that can reprogram and assemble themselves without external machinery. Here, the 3D printer is embedded in every particle of the material itself, so to speak – a material that has the inherent ability to perform information processing. This would be the ultimate manifestation of Chris Anderson's famous maxim "Atoms are the new bits."

G
Hacking the Process

→ #machines   → #materials   → #open source

The first creative wave in 3D printing used the then new industrial additive manufacturing technology available in high-tech centers as a tool for creating unprecedented three-dimensional forms. People had to play by the rules established by these machines and by all means had to avoid breaking them. The driver for the second creative wave is the increased accessibility of 3D printing technology thanks to online print services and, perhaps more important, the advent of open-source 3D printers. These low-cost machines essentially come supplied with a "license" to break rules or more precisely an invitation to modify and adapt the technology. This has led to a shift away from 3D printing as a tool towards 3D printing as a medium, encouraging the creative investigation of the tools and processes itself.

Hacking the Machine

Arduino is the poster child for the open hardware movement. Open hardware is technology designed according to the principles of open source culture, which calls for the free sharing of all the documents and blueprints necessary to build a piece of technology. The size of a credit card, Arduino is a low-cost electronics prototyping platform that features a microprocessor, or a tiny computer like those found in many household appliances such as microwaves. The micro controller is meant to talk with the world and is fitted with numerous input and output ports that enable the connection of sensors and actuators like motors and lights. The Arduino hardware is augmented with a straightforward programming language that lifts the barriers

to advanced physical computing. The platform is aimed at artists, designers, hobbyists, and anyone interested in creating rich interactive objects and environments. Since its inception in 2005 at the Interaction Design Institute Ivrea in Italy, Arduino has inspired thousands of creative applications and functions as an essential building block, the brains of many new open hardware projects.

One such creative offshoot is the RepRap open-source 3D printing platform, which adopted

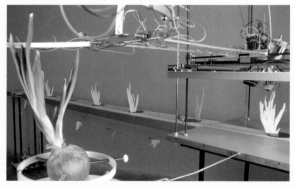

David Bowen, *Growth Modeling Device* → **p. 160**

the Arduino as its brains at an early stage of its community-led development. On this basis, it was only necessary to develop the electronics specific to the 3D printer, which piggyback on the Arduino as a so-called shield.

There is no singular model of open-source 3D printer. This technology can be viewed as a platform consisting of several elements, such as the positioning robot, the electronic brains, the printheads, and the software toolchains. Taken together, these building blocks form a 3D printer. A plethora of options exist for each of these elements, and these are often interchangeable. The world of open-source 3D printing is thus an ecosystem of modular building blocks, and just like Arduino, RepRap also inspires creative artists and designers to create new 3D printing processes and machines or to embed 3D printing technologies in their larger installations.

One of the first artists who incorporated open-source 3D printing in a kinetic sculptural piece was David Bowen with his Growth Modeling Device. At the heart of the contraption is an onion plant that is scanned at 24-hour intervals. As the scan slowly progresses, a 3D printer creates a plastic snapshot in real time. After each scan and print cycle, a conveyor belt moves the finished model and the cycle repeats, resulting in a series of plastic models that chronicle the growth of the plant. The printer Bowen used in his installation is self-built and based on the first RepRap model. To build the machine, he sourced components form various suppliers in the open-source 3D printer cottage industry.

The Solar Sinter, a machine that melts desert sand using focused solar energy in a similar way to industrial selective laser sintering, was built almost entirely from scratch by designer Markus Kayser. Keen observers familiar with the open-source 3D printing ecosystem will be able to spot several pieces of RepRap electronics in a documentary covering Kayser's trip to the Egyptian desert. Although creativity and lateral thinking are still the essential ingredients for new projects, open-source 3D printing building blocks can act as a stepping stone to something new.

Markus Kayser, *SolarSinter* → **p. 188**

Rael San Fratello's Emerging Objects platform researches 3D printing with cement, clay, bone, sand, and other unusual materials

Rock, Paper, Salt:
Imagining New Processes and Materials

Bowen's Growth Modeling Device was part of the 2010 exhibition Design by Performance at the Z33 House for Contemporary Art in Belgium. Also part of the exhibition, which focused not on the finished product, but on the performative nature of the production process itself, was l'Artisan Électronique. It was Unfold's first work that made use of the ceramic 3D printing process pioneered by the same team in the previous year. Based again on a 3D printer frame derived from RepRap, this time made by Bits from Bytes, Unfold replaced the standard printhead for thermoplastics with one they developed that extrudes tiny coils of clay.

The process has a great resonance with the age-old technique of coiling, in which a potter stacks coils of clay on top of each other – one of the first methods developed in Neolithic cultures for making utilitarian objects. Unfold makes all its ceramic 3D printing methods and printheads open source for others to build on. In search for ways to connect the physicality of a real pot with the virtual environment he liked working in, British potter Jonathan Keep was one of the first to adopt Unfold's method for printing ceramics in his own practice. He refers to the process as the fourth method of production in ceramics after hand building, wheel throwing, and molding techniques like slip casting.

Much of the research into printing with new materials is driven by artists and designers who are looking for materials that have better tactile and visual qualities, and a rich cultural heritage in their respective fields as opposed to the engineering plastics typically found in additive manufacturing. It is not only open-source machines that are used by creatives to explore new print materials: various groups use decommissioned industrial 3D printers with an expired warranty and load them with a plentitude of material. A popular machine to hack in this way is the ZPrinter from Z Corporation, which uses an inkjet powder printing process. Traditionally, this machine builds models out of a gypsum powder that is bonded using a binder sprayed from an inkjet head. But by loading different powders and experimenting with various binders, totally new materials can be printed like glass, paper, salt, and more ceramics.

Emerging Objects is a research-based architecture practice founded by architect duo Ronald Rael and Virginia San Fratello. Fueled by an interest in the creation of 3D-printed architecture, their research focuses on the development of new sustainable, inexpensive, and locally sourced materials for large-scale 3D printing. Most of their recent experiments make use of the inkjet powder printing process used by the ZPrinter. They have successfully printed with powdered paper, salt, wood, and cement, often sourced from local factory waste streams as well as all materials with a cultural lineage in the history of architecture. Throughout history, buildings were usually erected with local materials available on or near the construction site: wood in Norway, granite in France, and so on. The 3D-printed Canal House currently under construction by DUS Architects in Amsterdam will certainly stand out in all its plastic glory and draw massive crowds, but it will mainly act as a center for discussions on the future of 3D-printed architecture and not so much as a building meant to last. The research done by Emerging Objects opens the door for 3D-printed architecture that fits in with our urban and cultural fabric while still allowing for radical new possibilities of form.

Tool Marks

As the recent popularity of accessible 3D printing technology has shown, there is a demand for open tools that can be adapted and honed to one's own liking. In his essay "Tools," originally published in 2000 in a book covering the work of LettError, a collective of two Dutch typographers, Jan Middendorp argues for the importance of creating customized software tools instead of relying on mass manufactured applications. Middendorp refers to the fact that a craftsman, the predecessor of the designer, was never completely satis-

fied with the off-the-shelf tools sold in shops. "They always had the tendency to personalize their tools, to appropriate them by honing them, converting them or expanding them," he writes. In the same text, he describes the phenomenon of the tool horizon. Middendorp refers to software, but hardware tools are also imposed upon designers as if they were a preset straitjacket. Digital tools provide enormous possibilities, but they are never endless: sooner or later, designers will be confronted with the limitations of what they can do with them. Open-source building blocks help artists and designers to create their own complex tools in a similar way to traditional craftsmen. As a result, they can break away from the predetermined mode of design dictated by digital technology. As such, creatives can seriously intervene in the production process, and therefore in the final design, which bears the hallmarks of the tools it was produced with.

One such designer who has always built his own manufacturing tools is Dirk Vander Kooij. He believes the development of tools and processes is as interesting as the new design itself, and that the two develop simultaneously in a constant process of honing and adjustment. Inspired by the size limitations of industrial 3D printers, Vander Kooij obtained a decommissioned robot arm from a Chinese car factory to install in his studio. Equipped with a plastic extruder, the robot was turned into a giant 3D printer capable of printing furniture and other interior products. But instead of viewing his printer as a machine that should be able to produce any shape thrown at it, Vander Kooij recognizes the intrinsic strengths and weaknesses of his tool. One of the reasons why industrial manufacturers do not produce large-scale 3D printers is because every doubling of size on all

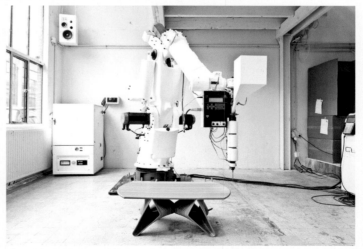

Dirk Vander Kooij, *Endless Table*

axes multiplies the print time by a factor of eight if the resolution and surface quality are retained. It is a traditional school of thought in 3D printing that it is better not to see the layers that make up the model – the tool marks. Not so for Vander Kooij and other designers who appreciate the imperfections of the production process, just like

Wobbleworks Inc., *3Doodler* → p. 78

Josh Bitelli, *Weld Drawings*

some potters leave the traces made by their hands in a thrown pot: "If you accept the low-resolution structure and try to treasure it rather than hide it, it can become a beautiful ornament."

The Aesthetics of Additive Fumbling

There is much debate as to what exactly constitutes 3D printing or additive manufacturing. What about computer-controlled knitting, for instance, where a thread of material is deposited layer by layer according to a digital blueprint? The additive concept of 3D printing has triggered a range of creative applications which might be considered out of bounds. Reintroducing the hand seems a popular trend in the highly mechanized world of additive manufacturing. Josh Bitelli produces steel vessels by manually drawing them layer by layer

with a MIG welding torch. Each successive layer is influenced by the small mistakes in the layer that came before it. The project was inspired by a conversation with an architect who mentioned that he could easily spot the standard shapes found in computer-aided design software.

A similar approach can be seen in Dave Hakkens's Precious Plastic project, in which the designer built a series of manual DIY manufacturing machines that process plastic waste. One of these machines extrudes molten plastic like the printhead in a fused deposition 3D printer, but in this case the models are again made by hand movements instead of machine movements. The fact that this is much more than a trip down memory lane in the world of craftsmanship is proven by the commercial merit of the 3Doodler, a printhead borrowed from a 3D printer that extrudes plastic, housed in a thick handheld enclosure that resembles a pen. The pen allows users to doodle in midair with lines drawn in solidified plastic.

Such innovations are inspired by processes that go beyond additive manufacturing, but each of these three projects completely ignores one of the important characteristics that defines additive manufacturing: the digital model. Developed by Joong Han Lee, Haptic Intelligentsia reflects on the age-old dilemma between the human hand and the machine. With this project, dubbed the "human 3D printing machine," Han Lee marries the imperfection and human determination brought by the combination of mind and hand with the perfection of a digital 3D model. Haptic Intelligentsia consists of a hot melt glue gun mounted onto a haptic feedback device. The device simulates force feedback: the user can move the glue gun freely, but once the tip of the gun is moved outside the surface boundaries of a virtual model, the system generates a force that pushes it back. The perceived resistance imitates the feeling of physically touching an object. Users are free to extrude the glue within the boundaries and to deliberately force the machine outside the predetermined rigid computer model.

The installation is a tongue-in-cheek reference to the origins of the FDM process, which was inspired by a glue gun, or, to use the affectionate nickname coined by Materialise founder Wilfried Vancraen, "fumbling deposition modeling."

H

Crafting New Industries

→ #empowerment   → #new craftsmanship   → #new industries   → #sustainable

Like so many of Istanbul's historic neighborhoods, the Şişhane district is full of shops that essentially stock the same items. While other areas are lined with shops selling musical instruments or kitchen supplies, the theme in Şişhane is lighting. At first, it seems strange that a few hundred similar businesses should cluster together in one area. Look closer and an ecosystem of entangled networks starts to unravel – one that benefits vendors and customers alike. On the surface one sees a couple of streets lined with shops selling finished lamps to private customers, but further down the winding alleys and in the anonymous buildings full of even smaller independent workshops, one can see and hear the machinery that makes Şişhane tick. Most people in the neighborhood do not actually sell to the public, but supply each other with all the little bits and pieces that make up a lamp. There are shops specializing in brass ornaments, workshops that create the graphics for lampshades, and sole traders who produce the metal frames the shades are stretched on. Right at the end of this chain are the retailers who sell lamps, and if a customer wants a custom-made chandelier, the individual preferences then trickle down the chain in face-to-face communication. Şişhane borders on one side to Galata, the hardware district, where the lamp makers source all their wires, knobs, and other electrical components, and so one cluster of craftsmen and sellers blends almost seamlessly into the next.

This method of working, in which very small businesses connect with each other and create a local ecosystem of their own, is no longer common, least of all in Western society. Most products consumed in the West have been mass produced in centralized factories and are then distributed on a global scale.

Looking at the bustling streets of Istanbul's historic center, we can recognize many of the characteristics at work which are often envisioned in future post-industrial manufacturing scenarios empowered by digital manufacturing and Internet communication – the so-called third industrial revolution. This is no coincidence, for an important segment of this future economy involves a partial return to the pre-industrial context of craft workshops, just like those found in the streets of Turkey which, until now, have managed to escape the centralized mass manufacturing models introduced by the previous industrial revolution. The wheel comes full circle.

A Brief History on the Making of Things

In the late eighteenth century, humanity entered the Industrial Age. With early inventions such as the spinning jenny, a spinning wheel that could produce multiple yarns at the same time and yield eight or ten times the output of a single spindle, the seed of industrial thinking was planted. The textile industry became the first to mechanize. By the second half of the nineteenth century, mechanization had spread to even more industries, and mechanical systems had replaced the need for manual labor in countless areas. When Ford introduced the assembly line in the early twentieth century and established a system of division of labor, efficient mass manufacturing became the norm for much of the previous century. This is now known as the second industrial revolution, and it is still the prevailing paradigm of manufacturing today. Products, buildings, and even entire cities are designed for the masses and defined by industrial standards.

Before the Industrial Revolution, most products were individually handmade by craftsmen who controlled the entire production process and had an extraordinary knowledge of all the materials and tools needed to create their items. This artisanal knowledge was often passed on from master to apprentice, and it took years if not generations to master. However, waves of industrialization began to erode this skilled craft knowledge and divided the making of things into many dis-

Unfold, *Stratigraphic Manufactury* → **p. 142**

crete steps that required little skill of those who performed them. It was no longer the actual maker who had an overview of the entire production of an artifact, but a legion of engineers and businessmen.

Looking at our manufacturing industry today, it is clear that inventors, designers, and creators do not have the power to produce and distribute the things they envision. In order to transform their ideas into products, they have to pitch to manufacturers and convince them of the merits of their design in the hope of getting it produced. Manufacturers may or may not take the risk of realizing a concept; there are more than a handful of middlemen such as marketeers, bookkeepers, investors, and engineers who all have potential reasons not to take the idea to market. And

not without reason, since the economies of scale require that a product be produced in enormous quantities in order to compensate the high costs of starting up production. Bringing a new product to market entails considerable risk.

Craft Meets Industry

With manufacturing going digital, we can now see aspects of the pre-industrial craft economy merge with high-tech industrial production methods. Such an approach will allow the production of things in much smaller numbers, in agile and well-networked microfactories located close to the users. One of the most important enablers and drivers of this shift is another digital invention which, in just 20 years, has managed to fundamentally change almost everything we do, from commerce to culture: the Internet.

In the context of those Istanbul streets, Unfold unveiled their Stratigraphic Manufactury project as part of the inaugural Istanbul Design Biennial. Stratigraphic Manufactury visualizes "a new model for the distribution and digital manufacture of porcelain, which includes local small manufacturing units that are globally connected. One that embraces local production variations and influences." What happens if no physical product is distributed, but only a digital file? How will this digital file be materialized in different local workshops, and how can the hand of the maker become visible again in a digital production process that is normally so uniform, exact, and sterile? A set of digital 3D designs was e-mailed to various makers around the world who were already acquainted with the ceramic 3D printing process that was required to produce the porcelain objects. The local manufacturers, most of them skilled ceramists, were asked not to modify the digital designs but were free to incorporate their personal signature and any local influences during the production and finishing of the pieces. This input ranges from the clay used – preferably local – to the glazes and finishes, but also the symbiosis between the craftsman and his personal machine.

Dirk Vander Kooij, Chubby Chairs

New Craftsmanship

Dutch designer Dirk Vander Kooij is a true craftsman, although most of the things he produces are actually not made by hand. He is a new kind of craftsman, one who embraces technology and machines, but not in the same way that industrial producers embraced their manufacturing lines. Dirk Vander Kooij is not interested in standardization or the mass production of identical items. Instead, he is interested in finding ways to create the industrial-quality products he likes AND selling them without having a rigid production system, upfront investments in tools and long production chains full of middlemen standing in the way. In short, he wants to design, create, improve, and create again.

Vander Kooij started off by creating his own large-scale 3D printer, a tool he immediately put to work by producing chairs and other furniture. A comparison between his plastic furniture and items of the same kind produced in a traditional process reveals some immediate differences. Plastic furniture such as the famous stackable Monobloc chair is industrially produced using injection molding. This technique is very well suited for such objects produced in large volumes. Due to the economy of scale, they are also cheap and, most of all, every chair is identical.

Injection molding entails forcing molten plastic into molds in order to make products. These molds are the most expensive part of the process. Once a mold has been created, though, it can be used to reproduce countless chairs. The more chairs produced and sold, the lower the production costs will be. Although the technique is very suitable for mass production, it does not allow any change. In other words, once the molds are made, the design is rigid and finished.

Vander Kooij's 3D printing robot is equipped with a very similar plastic melting device as an injection molding machine. However, instead of injecting the plastic into a mold, the material is deposited in an additive manufacturing process using a virtual mold in the computer memory. At the start of every new print cycle, the model can be revised without any need to invest in new molds or tools. This enables the designer to make improvements on every chair he prints, just like a craftsman would do as he acquires more advanced skills through repetition, making the design process fluid and flexible. Since he began producing his Endless Rocking Chair, Vander Kooij and his team have already created 54 iterations of the model. Every new print functions as the prototype of the next, but at the same time it is also a marketable product that will end up in stores, and not a mere prototype in the traditional industrial sense. Products thus do not have to be perfect from the beginning, unlike in traditional industrial manufacturing, because they can evolve over time, incorporating the criticism and recommendations of users.

This approach is radically different from how traditional industrial manufacturing used to work, and it is something of a return to the iterative process in which a craftsman used to constantly improve his wares. It is also closely related to the popular "release early, release often" philosophy in software development. The emphasis lies on the early release of a software product which is then constantly updated to iron out problems and to add or enhance features. This creates a feedback loop between the developer and users. It allows the developer to be more agile, while also involving the user in the development process, resulting in a better final product. It is not surprising that the digitization of manufacturing means we can already observe objects behaving like software.

Online Communal Value

Unlike the development of software, which does not need more than a laptop, time, and dedication, the manufacturing of goods will always require some amount of investment for tools and material. Because digital manufacturing can be scaled up from a single print, this investment can be kept small. There is no need to invest in molds, and there is no minimum production run or inventory because products are produced on demand, eliminating the need for storage. If a larger investment is needed to create a product, today's craftsmen have a trick up their sleeve that their ancestors did not: they can tap into a global network of investors using the Internet just as easily as it is for individuals to publish and promote their work to a worldwide audience using the same technology.

There are several ways to source funding for projects online. Most rely on the cumulative impact of many small financial transactions from a large number of people. Communication technology and online banking facilitate these micro transactions as they make it easy to send and receive money around the globe securely. Online crowdfunding platforms like Kickstarter, Indiegogo, Ulule, and Fundable are facilitators that connect people with an idea to other who believe in the same idea and are willing to fund it. These platforms can create a market before the product exists and an involved community. Backers, as funders are called, can invest small amounts of money in whatever projects they like. In exchange, they receive a gift from the creators related to the project. Depending on the pledged amount, this may be a limited edition of the product itself.

Engaging directly with an audience through online communications can bring many more benefits than funding alone. Feedback, ideas, help, improvements, and various types of information can be as valuable as money. Crowdfunding is thus an investment that is equally social and financial. Such communal pooling of funds and content alters the relationships between people. It encourages partnerships instead of unidirectional relationships based on personal interest, and this can foster a strong sense of community.

Microfactories and Green Energy

According to Jeremy Rifkin, author of *The Third Industrial Revolution: How Lateral Power Is Transforming Energy, the Economy, and the World*, we are undergoing a global transition. Traditional industry and economical systems as we know them, based on fossil fuels and a centralized system of production and distribution, are about to collapse due to the scarcity of natural resources. According to Rifkin, it is time to make the transition to a durable, green economy. Power should be obtained from decentralized sources that are more environmentally compatible, and every building should collect energy, feeding it into a local transmission network that offsets shortages in one area with surpluses in another.

Where Rifkin predicts a decentralized energy system, he foresees the same for production. Instead of big, centralized factories that produce in bulk, he proposes a decentralized, networked model with small production units that can meet local needs. Rifkin is not referring to the kind of local manufacturing networks seen in Istanbul, but to microfactories that could be located anywhere in the world, connected to each other through the Internet.

Markus Kayser's Solar Sinter is an extreme illustration of how digital manufacturing, local materials, and green energy can come together. He designed a 3D printer for use in a desert region which takes a locally abundant material (sand) and melts it using a locally abundant energy source (the sun) in order to produce utilitarian goods made of glass.

Rifkin has predicted a change from local decentralized production (craft), through global centralized production (industry) to a future in which a global decentralized production system becomes the new standard in manufacturing. Craft-scale production that utilizes industrial technology, facilitated by online networks.

Before we get carried away into this bright and shiny future, there are many hurdles to overcome. As Joseph Grima, curator of the 2012 Istanbul Design Biennial, states "the struggle to define new power structures, new economic frameworks, new forms of authority, new modalities of being political – an entirely new *social anatomy*, in other words – is unfolding in front of us at this very moment." In short, there can be no revolution without disruption.

3

→ Case Studies

Unfold and Tim Knapen
L'Artisan Électronique

→ #interact → #materials → #machines

Digital Craftsmanship with Amateur Hands

Year: 2010
Client/Purpose: For exhibition
Partners: Z33, Bits from Bytes
Technique: Paste extrusion (ceramic)
Printer/Service: Modified Bits from Bytes
Material: Ceramic, mixed media
Production time: N/A
Budget: $4,000
Edition: 1

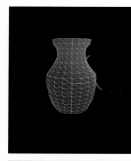

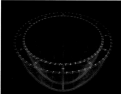

While industry and craftsmanship are positioned as polar opposites, they would be more accurately represented as volatile points in a matrix of manual, mechanical, and electrical forces. Wheel-thrown pottery, for example, though now considered an artisanal skill, developed as a partial automation of coil pottery by the third millennium BCE, rendering the production of small clay vessels more efficient. If industry is characterized by the displacement of advanced operations from human hands to machines, then handicraft is defined by the retention of fine motor skills mastered over years of practice. In L'Artisan Électronique, Belgian designers Unfold and Tim Knapen investigate the convergence of craft, industry, and digital making, defying straightforward categorization. The project is set up as a miniature production center, featuring a digital potter's wheel connected to a 3D ceramic printer. As a virtual cylinder spins on a computer screen, the user cuts away and elaborates its shape by passing his or her hand through a laser. When satisfied with the final form, the user can submit the customized model to a digital archive, which then supplies the 3D printer with geometric instructions. Finally, the printer traces the desired shape, layer by layer, in a process akin to coiling. By interweaving a variety of skillsets, L'Artisan Électronique removes making from the exclusive realm of the machine or the craftsman and introduces the amateur as an active contributor.

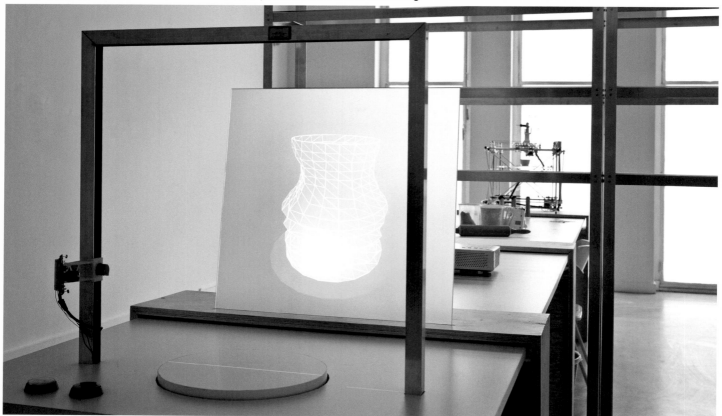

The shape of the digital vase is modified by passing a hand through the laser

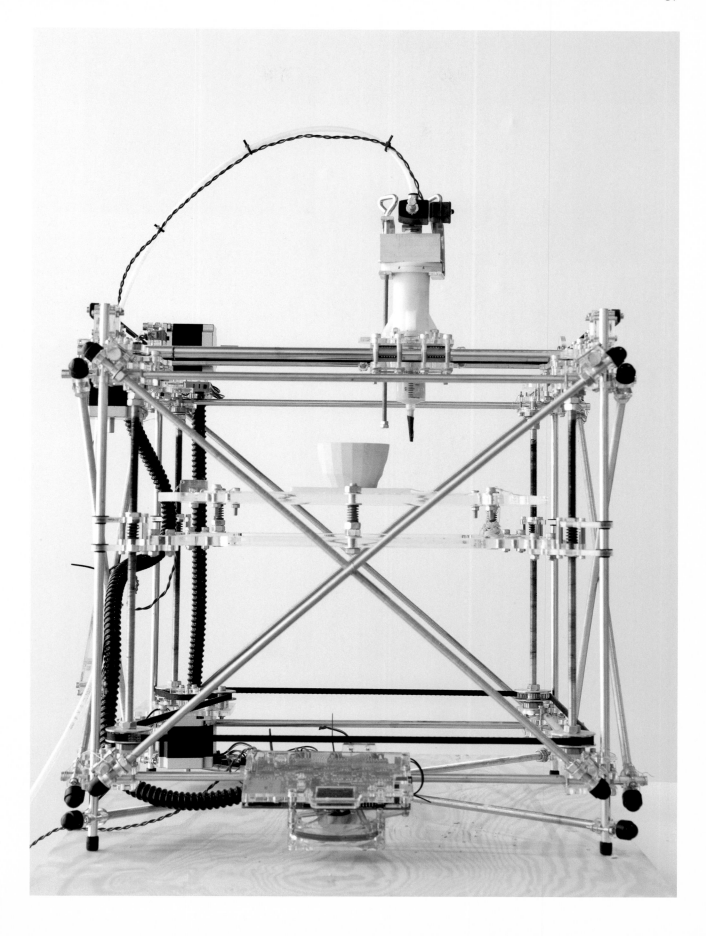

Different iterations are printed from the digital archive

Barnabé Fillion in collaboration with Unfold and Perla Valtierra

The Peddler

Year: 2013
Client/Purpose: For exhibition
Partners: Barnabé Fillion, Unfold, Perla Valtierra
Technique: Ceramic deposition modeling
Printer/Service: Modified Bits from Bytes Rapman
Material: Ceramic, oak, aluminium, leather, glass
Production time: N/A
Budget: N/A
Edition: 1

→ #complexity

Absorbing and Releasing Perfume in 3D-Printed Clay

While ceramics have been the source of many innovations in many high-tech industries, from aviation to biomedical implants, there remains a large disconnect between its treatment as a technically performative material and its treatment as a medium for craftsmanship. The technique for ceramic printing developed by the Belgian studio Unfold has begun to bridge the gap between the two poles. After first demonstrating the technique in 2010 with L'Artisan Electronique, the designers spent the following several years perfecting and disseminating the knowledge they had gleaned through the production of small vessels. The Peddler represents another evolution of their work, where the qualities particular to 3D printing (such as intricacy of structure and layering) are exploit-

ed for their associated performative effects. In collaboration with French perfumer Barnabé Fillion and Mexican ceramist Perla Valtierra, Unfold has created a set of 3D-printed ceramic tools, including containers and funnels for perfume, alcohol, and water, as well as an unglazed diffuser, whose complex pattern of cavities slowly releases the smell of perfume over an extended period of time. The ceramic printing layer height ranges from half a millimeter to one millimeter in thickness, carefully calibrated to the purpose of each object. A wooden structure allows for the circulation of the tools through the air for enhanced emanation of the scent.

A wooden structure for hanging the diffusers circulates the perfume through the air

Vessels of various shapes hold water, alcohol, and perfume for customized blends and intensities

etc. etc. design studio
Winter

→ #new craftmanship

A Family of Vases Undulates in Complex Facets

Year: 2012
Client/Purpose: Student research
Partners: Royal Danish Academy of Design
Technique: Paste extrusion (ceramic)
Printer/Service: Bits from Bytes Rapman
Material: Earthenware, stoneware, porcelain
Production time: 2 months
Budget: N/A
Edition: N/A

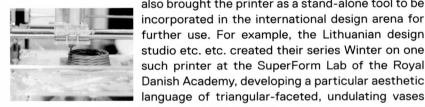

Following the diffusion of Belgian studio Unfold's ceramic 3D printing prototype in 2009, the technique has attracted more experimentation across a large global network. While the studio's Stratigraphic Manufactury project acts more as a collective game with its own set of rules, they have also brought the printer as a stand-alone tool to be incorporated in the international design arena for further use. For example, the Lithuanian design studio etc. etc. created their series Winter on one such printer at the SuperForm Lab of the Royal Danish Academy, developing a particular aesthetic language of triangular-faceted, undulating vases in order to reap the most benefits from the potential afforded by the machine. Their experiments with materials (porcelain, earthenware, and stoneware) and intermediary tools (such as a hairdryer) allowed them to hone the effects achieved with the volatile ceramic material, mixing the intuitions of a craftsman with the algorithmic approach of a technician. The family of vases, each varying in height and shape, is finished in a more traditional manner with a colorful glazed interior.

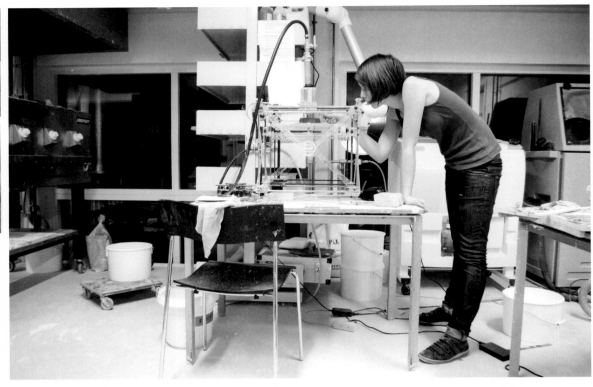

Experiments with different textural faceting

PARTYLab
Omote 3D Shashin Kan

→ #body topology   → #3D scanning

A Photo Booth of Printed Miniature Figurines

Year: 2013
Client/Purpose: iJet
Partners: Rhizomatiks, Engine Film, data design
Technique: Inkjet powder printing
Printer/Service: iJet
Material: Colored gypsum
Production time: N/A
Budget: $200 – 300 per print
Edition: On-demand

Handheld scanner

When asked by iJet to create a project promoting one of their new 3D printers, PARTY creative lab in Tokyo began by reflecting on the nature of 3D printing as a form of social interaction in Japan. They observed that, while rapid prototyping was used in the business-to-business world when developing commercial objects, and while private use was on the rise in some communities based on the spread of the maker movement, there was little visibility for additive manufacturing as a business-to-consumer service – something that could produce a finished object in a reasonable amount of time for customer interaction. At the same time, they did not want to offer something that could already be produced; in other words, the service

should leverage the full potential for end-user customization. The result, Omote 3D Shashin Kan, adapts the popular photo booth model ("shashin kan" in Japanese) to produce three-dimensional printed figures in full color, measuring between four and eight inches in height. PARTY worked with 3D scanners to optimize the results, using customized data correction software to automate the process from scanning to printing and to create a viable on-demand product: after 15 minutes of scanning with handheld tools, the customer can have the model printed directly.

blablabLAB
Be Your Own Souvenir

→ #3D scanning → #interact → #body topology

Printing Pedestrians in Barcelona in Miniature

Year: 2011
Client/Purpose: Research
Partners: Arts Santa Mònica, Agència Catalana de la Joventut, and Transports Metropolitans de Barcelona
Technique: Fused deposition modeling
Printer/Service: Bits from Bytes RapMan
Material: ABS plastic, PLA bioplastic
Production time: 3 months
Budget: $2,000
Edition: On-demand

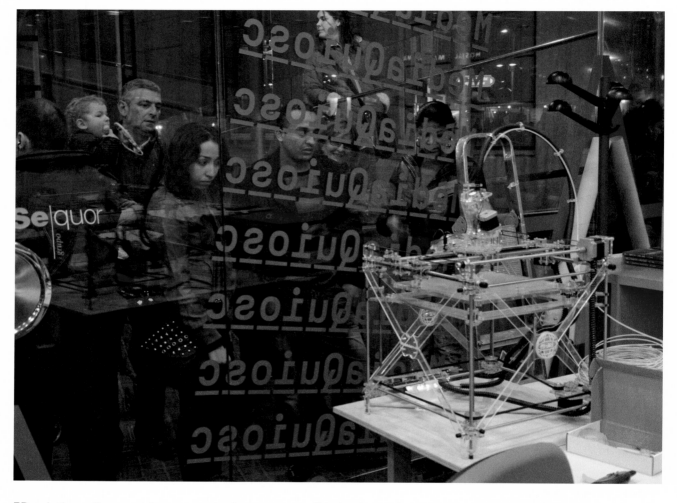

3D printing offers a unique opportunity to capture a precise moment in a physical object, like the concept of "spimes" developed by Bruce Sterling to describe objects that create an immediate link between digitally observable data and their physical manifestations. The Barcelona-based studio blablabLAB designed the Be Your Own Souvenir platform as an expression of the new possibility to create such objects using rapid prototyping and automated processes for modeling. Participants pose on a platform while three Microsoft Kinect sensors scan their body; a virtual model is generated from the information and then fed to a self-built Bits From Bytes printer, which prints a tiny figurine in a matter of minutes. By creating this installation in Barcelona's La Rambla neighborhood, blablabLAB have attempted to bring 3D printing out of the recesses of fab labs and industrial production centers into a public space, establishing a new form of exchange between producers and consumers that expands beyond the simple transaction of money. Furthermore, by referencing the idea of street performance and human statues, the project makes an implicit comment on the diversity of residents and agents who share the public space of a rapidly gentrifying zone in the city.

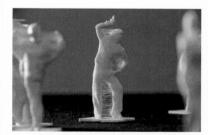

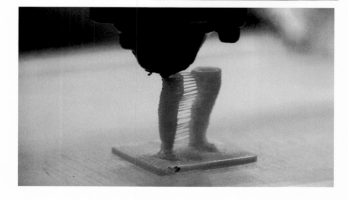

Axel Brechensbauer
Peace Drone

→ #building blocks

New Technology Puts on a Friendly Face

Year: 2013
Client/Purpose: Artwork
Partners: N/A
Technique: Selective laser sintering
Printer/Service: Shapeways
Material: Nylon
Production time: N/A
Budget: N/A
Edition: 2

3D printing – a loose assembly of related but highly variable practices – is a potent phrase in the collective consciousness of society today. We can sense and are finally beginning to test its potential to alter the way we relate to our material surroundings, while a thin voice of dissent raises questions about the consequences it may have on the environment or the labor and commodities market. Perhaps the military drone represents a parallel enigma, a modern technical device loaded with wild speculations, fear, and wonder. Swedish designer Axel Brechensbauer imagines a whimsical convergence of the two technologies with his peace drone, developed as an unsolicited proposal for the United States Army. The drone is designed to fly over foreign territories, emitting clouds of the narcotic painkiller oxycodone and broadcasting a cartoon-like smile from its 3D-printed head. As Brechensbauer puts it, "Happy people are better than dead people, and best of all they will be addicted to you!" – a fitting rejoinder to both the zealous devotees and scaremongers of our rapidly evolving technological infrastructure.

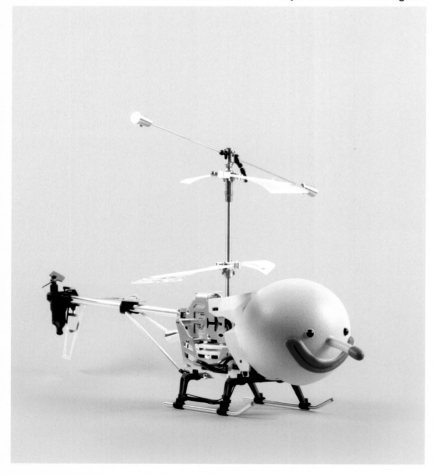

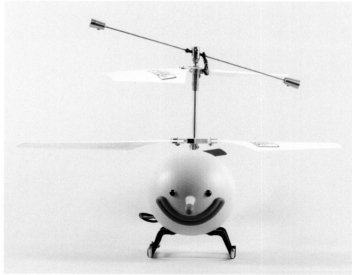

Andrew Spitz
loci

→ #customization   → #software

Visualizing Flightpaths with Interactive Software

Year: 2012
Client/Purpose: Student research
Partners: Copenhagen Institute of Interaction Design
Technique: Selective laser sintering
Printer/Service: Shapeways
Material: Nylon
Production time: 4 weeks
Budget: N/A
Edition: On-demand

The combination of economies of scale made possible by 3D printing is one factor that merges seamlessly with the rise in personalized data visualization. Projects like loci, developed by Andrew Spitz while studying at the Copenhagen Institute of Interaction Design, point to the evolving concept of memory, now increasingly realized as the creation of images or objects from statistical data (such as geolocation and flight bookings). Spitz's project creates mementoes for our contemporary form of travel, whose paperless tickets, e-passports, and border agreements have removed most of the physical remnants of a potentially significant personal experience. Each loci is generated by entering flight paths and digital tracking information on Foursquare or other media into a custom interface built using Max/MSP and later printed from a palette of diverse colors and materials using selective laser sintering. Form generation has been optimized to the extent that Spitz no longer has to make them personally; prints can simply be ordered from Shapeways. Looking ahead, he plans to integrate the creation of loci into Flying, an app he created as part of a development team at CIID.

Adam Nathaniel Furman
Identity Parade

→ #authorship

Mixed-Material Vessels Sample Contemporary Culture

Year: 2013
Client/Purpose: For exhibition
Partners: Design Museum (London), Lee3D, Figulo
Technique: Inkjet powder printing (ceramic), selective laser sintering
Printer/Service: 3D Systems ZPrinter
Material: Colored gypsum, nylon, resin, earthenware, porcelain
Production time: 3 months
Budget: $9,800
Edition: 10

Each year, London's Design Museum calls on four designers-in-residence to reflect on a theme, such as "identity" in 2013. Given the museum's focus on industry and everyday life, Adam Nathaniel Furman's Identity Parade was a surprising outcome. At first glance, this menagerie of luridly colorful and elaborate 3D-printed and slip-cast vessels recalls the avant-garde of gallery-based conceptual design. Yet Furman's work is rooted in the myriad obsessions of popular culture, from 4chan's LOLcats and YouTube's dancing Chihuahuas to Photoshopping, subliminal messaging, fluorescent animations, and old film clips removed from their original context, such as the atomic mushroom cloud over Hiroshima. For Furman, a graduate of London's Architectural Association, the transition between 3D modeling and laser sintering is no more belabored than the process of rendering an image of a virtual building design (just as

his multidisciplinary, speculative work is now a well-established mode of creative practice). With a cavalier pastiche of historical references that would make postmodernism look chaste, Identity Parade uses 3D printing to communicate our society's voracious, eclectic appetite for visual stimuli and our common disregard for the distinction between original and copy. As Furman says, "It is the act of copying or modifying itself that keeps the thing being copied alive [...] every reinterpretation forever alters the manner in which the thing being interpreted can be understood in the future."

WobbleWorks Inc.
3Doodler

→ #machines

The Power of Rapid Prototyping in a Handheld Pen

Year: 2013
Client/Purpose: For sale
Partners: Kickstarter
Technique: Fused deposition modeling
Printer/Service: 3Doodler
Material: ABS plastic, PLA plastic
Production time: N/A
Budget: $99 per piece
Edition: N/A

blooded exactitude of digital drawings. Launched in February 2013, the 3Doodler, a prototype for a handheld pen that extrudes plastic filament in midair, promises to combine the automation and material ease of 3D printing with the unlimited inventiveness of an ordinary pencil. Judging by its success on Kickstarter, where WobbleWorks were able to raise more than $1 million in just a few days, the 3Doodler has sparked the imagination of an audience that is much broader than the hardcore 3D printing community. Yet it also demonstrates how diverse the concept of additive manufacturing has grown: by removing the precise automation of CNC fabrication, the 3Doodler returns the extrusion process to the craftsman's sphere of fine motor skills.

As part of the design process, rapid prototyping may work best when used in a feedback loop, with continually testing new possibilities being refined rather than there being an aim for immediate perfection. However, the syntax of 3D modeling and the time-consuming printing process can compromise the spontaneity of the design method. This criticism has been levied against creative practice every time a technological development has threatened to replace a more analogue predecessor, from the demise of hand sketching brought about by AutoCAD to the introduction of drum machines. Though there is little proof behind such a claim, we still foster a belief in the unbridled *sprezzatura* or "studied carelessness" of the squiggly-lined napkin sketch over the cold-

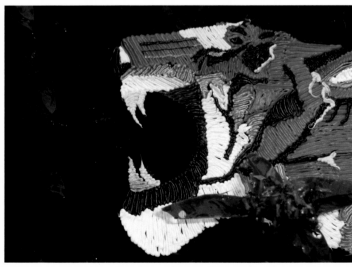

MAKIE Lab
Makies

→ #customization   → #new industries

Personalizing Dolls with Rapid Prototyping

Year: 2012
Client/Purpose: For sale
Partners: N/A
Technique: Selective laser sintering
Printer/Service: EOS Formiga P100, MakerBot Replicator 2, 3D Systems Cube
Material: Nylon, synthetic hair, acrylic eyes, fabric
Production time: 2 weeks per doll
Budget: $110 per print
Edition: On-demand

Entertainment has evolved significantly since the rise of smartphones and tablets, which enable the rapid diffusion of constantly evolving, niche games. As they require no physical disk or distribution system, games can often be sold at an extremely low price, and developers sometimes give them away for free. In most cases, however, a hybridized system of virtual and real currency is used to generate larger revenues, with costs assigned to functions such as unlocking levels and customizing avatars. One of the most successful models for children's games connects the digital environment to physical products such as stuffed animals, which may offer codes to augment the virtual game experience while also entering real-life playtime. MAKIE Lab, a London-based studio founded by Alice Taylor, has approached this concept in re-

Online user interface

verse, launching a novel platform to make customized dolls that will eventually tie in with an online game. Like many personalized dolls, the creation of each MAKIE involves selecting the skin tone, hair color, style, and outfit, However, the company goes much further by turning the precise shape of the eyes, ears, nose, lips, and jawline into user-defined parameters. The unique doll is then 3D-printed, accessorized with the selected hair and clothing, and delivered to its new owner. The site already has an active community forum that includes user galleries and instructions for hacking or modifying the doll. This may develop into a full-fledged social network once the game is released.

Cunicode
Crayon Creatures

→ #customization → #software

Physical Mementos from Children's Drawings

Year: 2013
Client/Purpose: For sale
Partners: N/A
Technique: Inkjet powder printing
Printer/Service: 3D Systems ZPrinter
Material: Colored gypsum
Production time: 3 weeks per piece
Budget: $130 per print
Edition: On-demand

Compared to other craft techniques, rapid prototyping may fail to spark the kind of emotional attachment that is often associated with flawed, timeworn, or self-made objects. Nevertheless, the ease with which most 3D-printed objects can be produced makes them suitable as keepsakes or mementos, especially given the possibility for serial customized production. Crayon Creatures, a project by Bernat Cuni at the cunicode studio in Barcelona, began life as a simple personal project for the designer's daughter, but it has since evolved into a small-scale business platform for creating similar objects on demand. Parents can send in their children's drawings, and cunicode will transform the colorful figures into "inflated" three-dimensional models using texture mapping to reproduce the visual language (Cuni tends to identify the front or side of a head to orient the drawing). Next, the image is sent to a 3D printing service such as the Dutch–American company Shapeways or the Belgian Materialise, where the figurine is printed in a full-color, gypsum-based powder. Finally, the object is delivered to the customer's home. Crayon Creatures demonstrates the low entry barrier to customized manufacture thanks to a network of industry partners connected by the Internet.

AMINIMAL/John Briscella
INKIMALS

→ #customization → #software

Hand-Drawn Toys Printed in Full Color

Year: 2013
Client/Purpose: For sale
Partners: N/A
Technique: Inkjet Powder Printing
Printer/Service: 3D Systems ProJet 660Pro
Material: VisiJet PXL
Production time: N/A
Budget: N/A
Edition: On-demand

As full-color printing becomes more ubiquitous, it is also rapidly increasing the reach and viability of customizable products as a commercial design strategy. A 3D model is still difficult for most people to conceptualize and alter, let alone build from scratch; similarly, a two-dimensional drawing is not something that can be automatically translated into a solid virtual model without a considerable amount of arbitrary (and thus laborious) interpretation on the part of a more experienced designer. Thus, for the most common types of 3D printing using a single material, the degree of personalization for end users has generally been limited by their fluency with 3D modeling. Coloring, on the other hand, is both easy to do and simple to apply virtually to an existing shape. John Briscella at the Aminimal design studio has capitalized on this distinction by launching inkimals, a platform for creating customized toys. He offers the user a variety of simple anthropomorphic shapes, rendered as two blank templates (front and back). Users can then create whatever appearance they like by coloring inside the lines and submitting a digital photo. This photo is then mapped directly onto the virtual model and 3D-printed in full color, replicating every detail of the original sketch.

INKIMALS come in a variety of basic three-dimensional shapes, customized by coloring in flat outlines

Tecnificio
Stampomatica

→ #building blocks

Reinventing Letterpress for Distributed Self-Production

Year: 2013
Client/Purpose: Research
Partners: Lino's Type
Technique: Fused deposition modeling, laser cutting
Printer/Service: KentStrapper Volta, CO₂ Laser Cut 350W
Material: ABS plastic, PLA bioplastic, fiberboard
Production time: 4 months
Budget: $20,000
Edition: In development

While 3D printing is transforming the field of industrial manufacturing, a popular nostalgia has arisen for older mechanical techniques dating back to the birth of the industrial era. The aim at the time was to achieve more efficient standardization, but contemporary audiences now delight in the imperfect products of automated but analog methods, such as vinyl records and Polaroid snapshots as compared to identical MP3s or rather unromantic digital photos. Stampomatica, on the other hand, represents a fusion of the best of both: the distributed, accessible, and adaptable nature of rapid prototyping with the personalization and craftsmanship of analog tools. Emerging from a collaboration between the Tecnificio production facility outside Milan and a handmade letterpress called Lino's Type from Verona, the small machine is designed for the at-home production of small printed paper goods such as business cards and postcards. The Stampomatica apparatus combines laser-cut fiberboard parts with 3D-printed clichés, or printing plates. Unlike the more specialized and laborious technique of metal casting, the plate can be manufactured on any 3D printer or ordered online; the creators are also planning an online service to facilitate the process. The project suggests an interesting model for the small-scale resurrection of industries that have otherwise been endangered since the advent of digital technologies, while expanding the domain of these crafts from experienced professionals to amateurs and dilettantes.

Stampomatica combines a laser-cut structure with 3D-printed printing plates

Léo Marius
Open Reflex

→ #empowerment → #building blocks → #open source

An Open-Source Camera for Desktop Printing

Year: 2013
Client/Purpose: Student research
Partners: École Supérieure d'Art et Design Saint-Étienne, RandomLab
Technique: Fused deposition modeling
Printer/Service: MakerBot Replicator 2X
Material: ABS plastic
Production time: 3 months
Budget: $40 per print
Edition: In development

Although 3D printing is often discussed as if it comprised a monolithic apparatus of production, the actual differences in terms of quality, cost, and accessibility between high-end models and the basic RepRap is so great that most general observations are meaningless. The notion of everyone being able to produce functional consumer objects on a home 3D printer is still mostly incompatible with the potential of low-cost personal printers. For Léo Marius, however, the limitations of such a machine became a fruitful challenge. In his final project for the École Supérieure d'Art et Design Saint-Étienne, he designed an analog camera with a mechanical shutter that could be printed in under 15 hours on a common 3D printer (in this case, the MakerBot Replicator 2X), accommodating any photographic lens with an adaptable mount ring. Furthermore, the Open Reflex was designed completely on open-source 3D modeling software and then uploaded to Thingiverse so that anyone is able to customize the schematics. So far, such modifications have ranged from an Arduino-powered test bench shutter to measure shutter speed to a homemade camera lens designed by Yuki Suzuki. For the time being, the Open Reflex is not available for sale; those interested must produce it themselves, increasing the possibility of crowdsourced improvement.

Joris van Tubergen

€1,- per minute design

→ #new industries → #open source

Reducing Design to the Bottom Dollar

Year: 2011 to present
Client/Purpose: Research
Partners: N/A
Technique: Fused deposition modeling
Printer/Service: Ultimaker Original
Material: PLA bioplastic
Production time: Varied
Budget: $27,500
Edition: On-demand

Van Tubergen's entire production facility is easy to move

In a 1903 essay entitled "The Metropolis and Mental Life," German sociologist Georg Simmel hypothesized that the new urban condition of extreme population density would have several consequences: first, that human interactions would become valued exclusively in terms of easily divided, identical units of money and time; second, that the sheer volume of exchange would generate increasingly specialized objects of desire; and third, that the individual would tend towards exaggerated forms of visible signs of uniqueness. From this perspective, Dutch designer Joris van Tubergen's model of €1,- per minute design reflects the pressure to quickly produce a wide array of customizable objects on demand. In order to fit the timescale of a shopping experience, he has modified his Ultimaker 3D printer to deposit a thicker line of plastic with a higher layer height. Since all of the objects (bracelets, vases, cups) are manufactured in the same automated way, there is no qualitative valuation of different craft techniques, and the "work" can thus be priced solely in terms of printing time, hence the name of the project. Moreover, if a buyer desires a cheaper object, the speed can simply be adjusted to produce a faster print of lower quality. The fact that van Tubergen's

designs are all open-source, however, suggests a newer, more complex economical ecosystem for the modern era, in which intellectual property need not be privatized and may have another form of valuation altogether.

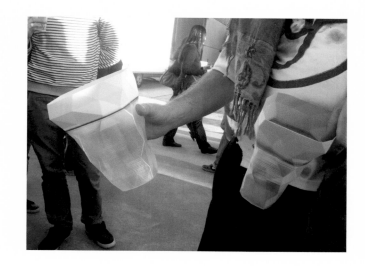

Kevin Spencer
Mini Designer Chairs

→ #authorship

Iconic Furniture Models Become Printed Copies

Year: 2013
Client/Purpose: For sale
Partners: Shapeways
Technique: Inkjet powder printing
Printer/Service: Shapeways
Material: Colored gypsum
Production time: N/A
Budget: $6 to $770 per print
Edition: On-demand

Gerrit Rietveld's *Red Blue Chair* **(1:12 scale)**

The furniture from the Swiss company Vitra, though it is beyond most people's budget, remains one of the most significant references for designers and architects alike. Such professionals often populate their renderings with 3D models of the iconic Eames Lounge Chair or Mies van der Rohe's Barcelona Chair to convey an immediate sense of refined living. Designer Kevin Spencer has now made a series of these models available as full-color, 3D-printed sandstone miniatures at a scale of 1:12 from Shapeways, the Dutch 3D printing service founded in 2007. Notably, such miniature chairs are already available from Vitra itself, crafted in painstakingly realistic detail (costing up to ten times the price of Spencer's designs). Spencer's ability to sell such direct derivations of copyrighted design pieces demonstrates the ambiguity of intellectual property in the realm of 3D printing, even when dealing with a company such as Vitra, which is known for quite vigorously pursuing the producers of unauthorized life-size furniture copies. On the other hand, Vitra has never posed any objection to the circulation of virtual models of its furniture, perhaps because this has functioned as a free form of advertising – and this digital availability may have made Spencer's work all the more effortless. At the same time, though, his products constitute a rather significant material transformation. For the time being, these tiny chairs occupy a gray zone of tentative fair use within the field of reproducible creative work.

Mies van der Rohe's *Barcelona Chair* **(1:12 scale)**

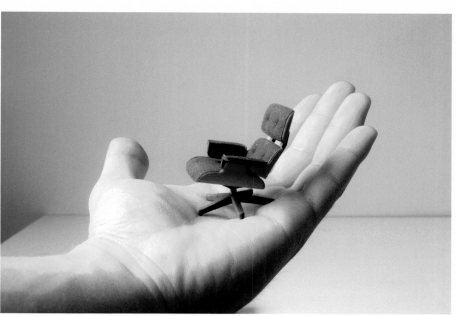

Charles and Ray Eames's *Lounge Chair* **(1:20 scale)**

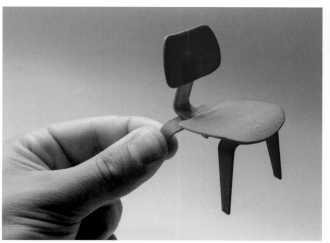

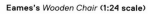

Eames's *Wooden Chair* **(1:24 scale)**

Samuel N. Bernier and Andreas Bhend
Draisienne IKEA Hack

→ #empowerment → #building blocks

A Mass-Produced Stool as a Customized Building Block

Year: 2012 to 2013
Client/Purpose: Research
Partners: Le FabShop, Instructables
Technique: Fused deposition modeling
Printer/Service: MakerBot Replicator 2
Material: ABS plastic
Production time: 5 hours per piece
Budget: $270
Edition: In development

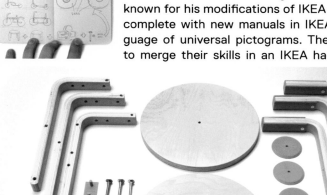

After exploring the use of 3D printing to interface with standardized objects in Project RE_, Paris-based designer Samuel N. Bernier was challenged by design blog core77's Ray Hu to collaborate with Andreas Bhend, a design student from Basel known for his modifications of IKEA furniture kits, complete with new manuals in IKEA's iconic language of universal pictograms. The two decided to merge their skills in an IKEA hacking project,

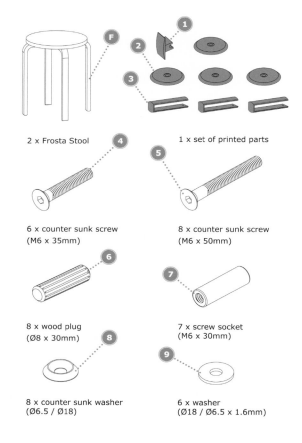

2 x Frosta Stool

1 x set of printed parts

6 x counter sunk screw (M6 x 35mm)

8 x counter sunk screw (M6 x 50mm)

8 x wood plug (Ø8 x 30mm)

7 x screw socket (M6 x 30mm)

8 x counter sunk washer (Ø6.5 / Ø18)

6 x washer (Ø18 / Ø6.5 x 1.6mm)

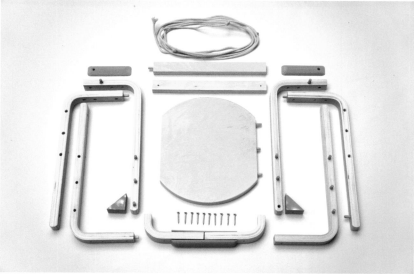

The standard IKEA FROSTA stool parts can be recombined to make different objects

using custom-printed items to transform the parts of a €10 FROSTA stool (inspired by Alvar Aalto's 1933 Stool 60 for Artek) into new objects. Over the course of several days, the two designed two products for children, including the DIY Luge sled and the Draisienne bicycle. The pair used only the parts taken from the IKEA stool kit or printed on a MakerBot 3D printer (shown in the orange hue of Project RE_), as well as a drill, pliers, and a metal saw. For the Draisienne, the round seats were used as wheels with the curved legs serving as the frame and handles of the bicycle. Both the Draisienne and the DIY Luge were later posted to Instructables so that anyone could assemble the toys for their children.

Matthew Plummer-Fernandez
sekuMoi Mecy

Year: 2012
Client/Purpose: Research
Partners: N/A
Technique: Inkjet powder printing
Printer/Service: Materialise
Material: Colored gypsum
Production time: 2 months
Budget: $160
Edition: 1

→ #authorship

The Friendly Face of Digital Contraband

Design at its most basic level, a process of assembling materials into functional objects, has been notoriously difficult to regulate as intellectual property: unless a specific scientific innovation can be proven, the exclusive rights to produce furniture must be proven ad infinitum as bootleg versions populate an international marketplace, seemingly unstoppable. Curiously, 3D printing (like music sharing) both supports and compromises the premise of artistic copyright; it is much easier to show that one source file is identical to another, but it is also much easier to circulate files via public hosting sites or peer-to-peer networks. Matthew Plummer-Fernandez, a digital process artist, operates at the ambiguous borders between private property, fair use, and Creative Commons licenses, but he has set himself a much more formidable challenge by targeting Disney, famous for the fiercely litigated protection of its characters. Plummer-Fernandez created sekuMoi Mecy as a project to test the limits of copyright: after 3D-scanning a 1972 rubber Mickey Mouse toy, he morphed and recolored the virtual model using software written in Processing. He then had the file printed by Materialise, who later refused to produce any more iterations of the model. While such corporate entities could easily be tracked and sued, it would be nearly impossible to monitor such output on personal 3D printers. Still, sekuMoi Mecy begins to trace the outlines of how much territory an idea or image can claim for itself.

Matthew Plummer-Fernandez
Smooth Operator

→ #authorship

The Fuzzy Borders of Abstracted Copyrights

Year: 2013
Client/Purpose: Research
Partners: N/A
Technique: Inkjet powder printing
Printer/Service: 3D Systems ZPrinter
Material: Colored gypsum
Production time: 2 weeks
Budget: $160
Edition: 1

After beginning to test the limits of copyright protection with his 2012 project sekuMoi Mecy, Matthew Plummer-Fernandez has continued to explore the thorny question of protected property in its manifold aspects, using the same Mickey Mouse toy produced by Disney in 1972. Plummer-Fernandez's work focuses on the paradoxical disparity between how we treat 3D files as opposed to printed objects: while the former are "more liquid than solid" and can be endlessly morphed, transformed, and distributed at virtually no cost, the physical counterpart – "the clumsy lump of plastic or dust," as he terms it – is the thing that stimulates reactions, be they the positive sentiment of ownership and care or the negative impulse to regulate and destroy. In his latest project, Plummer-Fernandez used a photogram of the toy (a 3D mesh built from a computer analysis of photographs from multiple angles) as the basis for

Smooth Operator, a software application he designed using Processing. This program creates a sliding scale of smoothing operations, both in terms of color and form, working like a three-dimensional version of the blur tool in programs like Photoshop. In contrast to the tendency for 3D printing to become more and more accurate in rendering fine details, Smooth Operator generates a highly abstracted result, but one that is useful for obscuring recognizable images to the point of vague reminiscence when their exact resemblance would incur liability.

Unfold
Kiosk

Year: 2011
Client/Purpose: For exhibition
Partners: Z33, Bits from Bytes, 4D Dynamics, Jo Van Bostraeten, OpenStructures, Mu, Polhemus

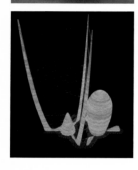

Technique: Fused deposition modeling
Printer/Service: Bits from Bytes BFB3000
Material: ABS plastic, PLA bioplastic, mixed media
Production time: N/A
Budget: $4,000
Edition: 1

→ #empowerment   → #authorship   → #new industries

Scanned Bootlegs on the Contemporary City Street

Counterfeit objects are a ubiquitous element of modern commercial enterprise, from the fake Casio watches and luxury handbags hawked on New York's Canal Street to the bootleg DVDs sold in the malls of Bangkok. With the rise of low-cost 3D printing, this offshoot of mainstream industry is set rise exponentially. With Kiosk, Unfold offers a theoretical speculation on the conditions for counterfeit production in the future, modeling the delivery model of a one-man sales unit on mobile street food carts. When the project was launched at the 2011 Salone del Mobile in Milan, Unfold used a 3D scanner on objects presented by various designers, immediately generating virtual models of their work. Later, driving around the city with the 3D printer and mobile shop onboard Kiosk, Unfold could 3D-print these poached models – or any others from a database of virtual iconic design pieces, ranging from Iittala's Aalto vase to Philippe Stark's Juicy Salif orange squeezer for Alessi – and sell them to passersby. They could also meet any requests for alterations in size, proportions, and color, as well as more extensive modifications, for a fraction of the price of the original object. Kiosk raises rhetorical (for the moment) questions about the terms of authorship and aesthetic control in the future, suggesting the need for designers themselves to define what position they wish to represent in a more ambiguous era.

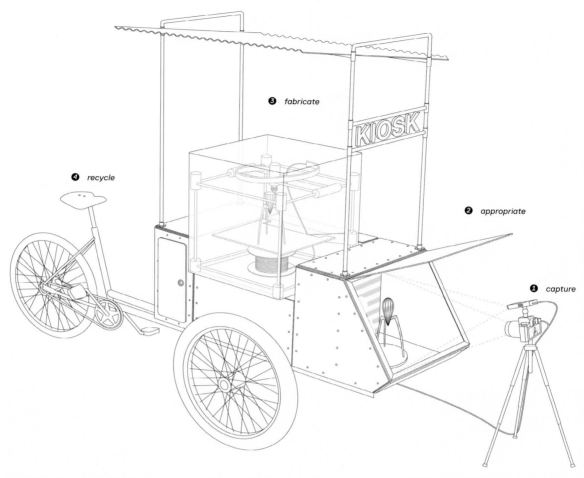

❸ *fabricate*

❹ *recycle*

❷ *appropriate*

❶ *capture*

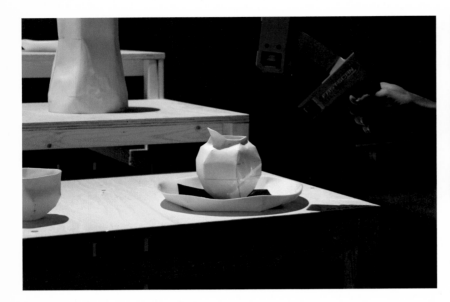

Work by other designers is 3D-scanned for reproduction at the exhibition After the Bit Rush at MU Eindhoven

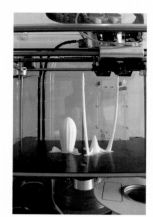

Weilun Tseng
Open E-Components

→ #empowerment　　→ #building blocks　　→ #open source

From Proprietary Appliances to Open-Source Tools

Year: 2013
Client/Purpose: Student research
Partners: Evan Frenkel, Marco Gambino, Minghan Tsai, Design Academy Eindhoven
Technique: Selective laser sintering, inkjet powder printing (ceramic)
Printer/Service: N/A
Material: Nylon, clay
Production time: 1 to 14 days per piece
Budget: $130 to $800 per print
Edition: In development

In his 1970 book *The Consumer Society: Myths and Structures*, Jean Baudrillard noted that the "general climatization of life, goods, objects, services, behaviors, and social relations represents the perfected, 'consummated' stage of evolution" – where the complete facilitation of everyday life is predicated on the desire to consume rather than any real functional necessity." When the conditions of modern life emerged from centralized systems of manufacture, this condition could hardly be challenged; today, however, the availability of information and customization may trigger a change. With Open E-Components, Taiwanese designer Weilun Tseng targets the domestic electronics market by dismantling its financially motivated complexity. After studying 50 consumer appliances, he reduced them to five simple actions: rotating, lighting, air heating, liquid heating, and direct heating. He then created a collection of five functional modules and 17 supporting parts (mostly 3D-printed with some standard elements) that could be combined to replace the myriad appliances currently sold with large markups. Tseng himself recreated a variety of commonplace objects, such as a table lamp, fan, milk frother, water boiler, and hair dryer, but he also opened the system to crowdsourced contributions, which reveal more niche configurations. These include Evan Frenkel's DIY slide projector, Marco Gambino's hybrid sunlamp and cactus planter, and Minghan Tsai's shoe dryer.

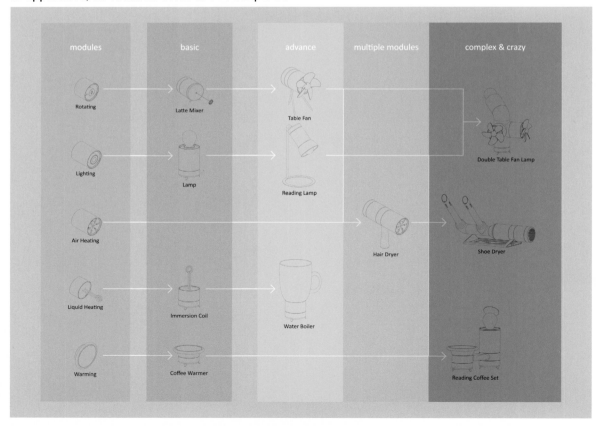

The five modules can be augmented with accessories and used in combination to produce a huge variety of functional objects

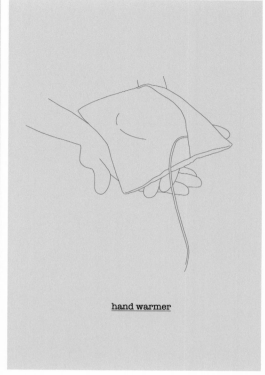

coffee heater

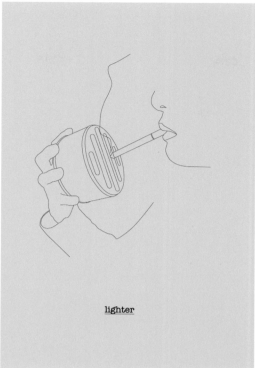

lighter

hand warmer

Different functions of the heating module

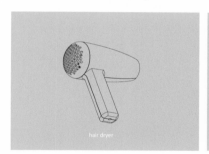

hair dryer

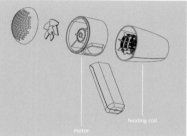

motor

heating coil

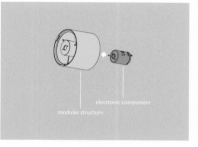

electronic component

modular structure

A normal hairdryer

An exploded axonometric shows the components hidden within the appliance

The Open E-Components hairdryer uses the heating and rotating modules

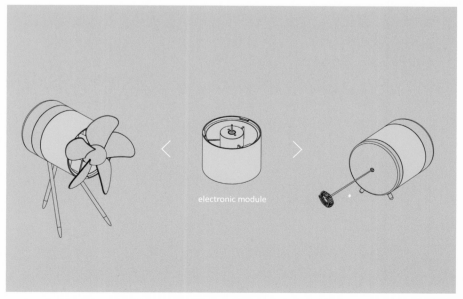

electronic module

One module opens up multiple options for new appliances

Electronic modules, from left to right: liquid heating module, lighting module, warming module, air heating module, rotating module

Lamp

Reading lamp

Table fan

Milk frother

Weilun Tseng

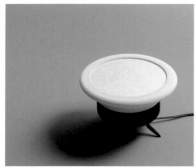

A hot plate warms a cup of coffee

Water boiler

Hairdryer

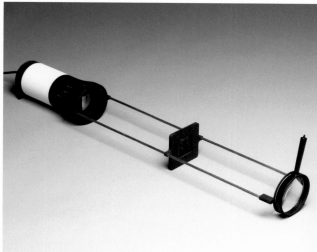

A DIY slide projector made by Evan Frenkel

Marco Gambino's sunlamp features cactuses and succulents around the bulb that thrive on the excess heat

Minghan Tsai's shoe dryer uses the components to circulate heated air through wet shoes

Jesse Howard
Transparent Tools

→ #empowerment   → #building blocks   → #open source   → #sustainable

Open Assemblies for Deconstructed Appliances

Year: 2012
Client/Purpose: Student research
Partners: Protospace Utrecht, Gerrit Rietveld Academie
Technique: Inkjet powder printing, fused deposition modeling
Printer/Service: 3D Systems ZPrinter, Ultimaker, MakerBot
Material: Gypsum, PLA bioplastic
Production time: 4 months
Budget: $1,000
Edition: On-demand

The fact that domestic appliances all tend to work in a few basic ways is often disguised by gratuitous functions (with associated leaps in price), proprietary casing, and intimidating technical instruction manuals that discourage repair. It follows that when they stop working or suffer exterior damage, they tend to be discarded and replaced. In his graduation project for Amsterdam's Rietveld Academy, Jesse Howard developed Transparent Tools as an alternative way for everyday consumers to buy and use electrical appliances, as well as a new definition for the working method of designers. Using additive manufacturing and the potential of networks, Howard assembled a series of household objects (a toaster, a coffee grinder, a vacuum cleaner, and a kettle) from 3D-printed parts, repurposed standardized objects, and salvaged components found on eBay. Each appliance

was then published as a single sheet of instructions, with an exploded diagram showing each component and where it could be found with URLs to virtual models uploaded on Thingiverse or search terms in eBay. The "transparency" of this design approach facilitates both repair and substantial modification: rather than pursuing an iconic visual form, Howard makes it clear that the design can be adapted in line with the specific context of any individual maker. For example, he improvised a second vacuum cleaner using a plastic thermos and expanded the toaster model to an industrial scale using the heating element from a ceramic kiln.

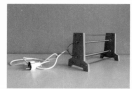

Howard improvised one vacuum cleaner using a plastic thermos

Samuel N. Bernier
*Project RE_*

→ #empowerment   → #building blocks   → #open source   → #sustainable

Helping Users Hack Their Object Cycles

Year: 2012
Client/Purpose: Student research
Partners: Instructables
Technique: Fused deposition modeling
Printer/Service: N/A
Material: ABS plastic
Production time: 3 months
Budget: $700
Edition: N/A

Despite the prevalent forecasts of a maker revolution, much of the output from low-cost, personal 3D printers could be classified as novelty objects, largely due to size constraints and material weakness. While completing his master's degree at the University of Montreal, designer Samuel Bernier decided to explore ways of using 3D printing to interface with abundant, standardized objects whose function had been superseded, like plastic bottles, tin cans, and glass jars. In partnership with Instructables.com, Bernier developed Project RE_ as a platform for open-source household objects made of empty containers and customized attachments produced on a widely available 3D printer through fused deposition modeling. The initial collection consisted of 14 objects, including a dumbbell, a piggy bank, a birdhouse, a lamp, and a snow globe. Bernier's approach recalls the work of Victor Papanek, who questioned the wasteful exploitation of conventional industrial design production chains for financial ends, and instead proposed simple alterations to existing objects and environments to satisfy basic human needs.

Lemon squeezer

Watering can

Mug

Piggy bank

Dave Hakkens
Precious Plastic

→ #new industries   → #machines   → #materials   → #sustainable

Recycling Plastic with Micro-Industries of Tolerant Machines

Year: 2013
Client/Purpose: Student research
Partners: Design Academy Eindhoven
Technique: Melting, injection molding, rotation molding, extrusion
Printer/Service: N/A
Material: Polypropylene, polyethylene
Production time: 6 months
Budget: N/A
Edition: In development

Shredded plastic

Extrusion process

Prototype plastic gun

The extrusion machine

Overview of the machines

Viewed through a longer historical lens, 3D printing is simply another phase in the manipulation of synthetic plastics over the past century, a process that is growing ever more freeform and precise as the technology evolves. While both our ability to manipulate plastic to meet our desires and our demand for high performance continue to grow, however, we treat the material, and especially its disposal, with a rather casual disregard. In fact, the extreme pressure to achieve efficiency and perfection in the manufacturing industry means that only 10% of the plastic we use is recycled, as Dave Hakkens discovered while working on his graduation project at Design Academy Eindhoven. He responded by devising a production model that could tolerate material variation and reuse. The Precious Plastic machines shred used plastic items into tiny pieces, which are then melted and reformed into new objects through different processes, such as extrusion and injection or rotation molding. Like Kieren Jones's Chicken Project, Precious Plastic cannibalizes and reappropriates old domestic appliances and industrial equipment to create small-scale factory setups, envisioning a future of decentralized manufacture in which the consumer actively contributes to the cycle. In the future, the Precious Plastic setup may be able to feed reused plastic directly into 3D printers, which have been notoriously unfriendly to recycled plastic, until very recently, in their drive towards optimization.

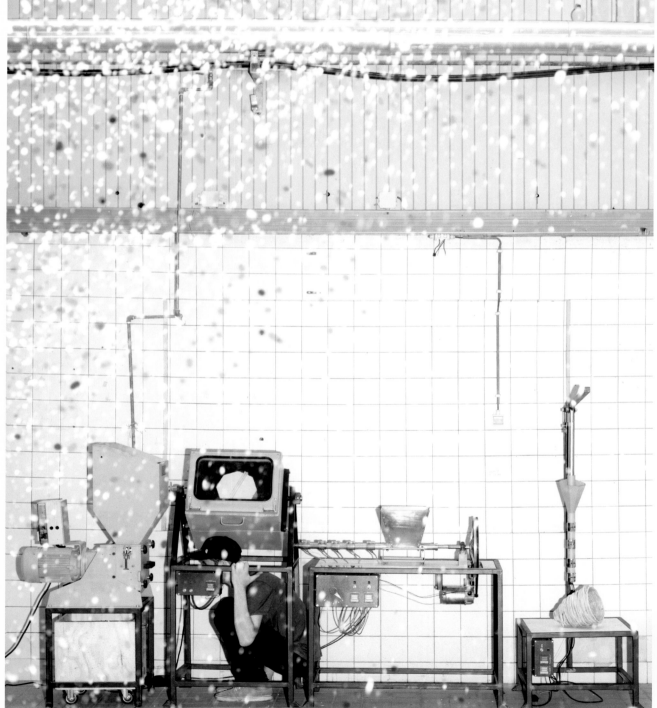

Extruded plastic

Spinning toys

David Graas
Screw It

Year: 2013
Client/Purpose: For sale
Partners: JP Melville
Technique: Selective laser sintering
Printer/Service: N/A
Material: Polyamide
Production time: 6 months
Budget: $30 to $140 per print
Edition: On-demand

→ #building blocks

Connecting Empty Bottles as Functional Structures

The standard bottle cap connection, modeled in 3D

Beyond the idea of perfect objects emerging fully formed from the 3D printing bed, rapid prototyping also allows for nuanced interventions, add-ons, and "prosthetics" for existing objects. These intermediary components take advantage of the high precision of digital modeling in order to interface seamlessly with standardized joints while minimizing the size of the part that must be 3D-printed. David Graas's Screw It collection, designed for the Dutch company Layers, works as a lattice into which the universal plastic drinking bottle can be screwed to create larger structures with different functions. For example, the bottle attaches to a 3D-printed wheeled chassis to make a toy truck, stores dry food for the kitchen in a balloon-dog configuration, or holds water and flowers in a modular vase. Meanwhile, Graas also adapts the bottle cap as a "jewel" to be screwed onto a 3D-printed ring or bracelet. Screw It employs a hybrid design language, finding a fruitful intersection between the worlds of mass manufacturing, which can be found almost everywhere in the world, and on-demand production, realized through distributed printing services.

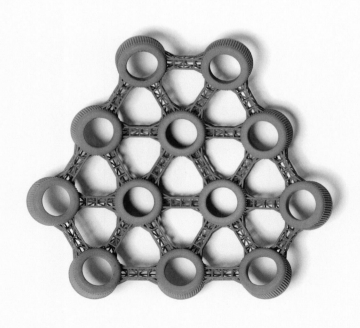

Minale Maeda
Keystones

Year: 2012
Client/Purpose: Research
Partners: N/A
Technique: Selective laser sintering
Printer/Service: Materialise
Material: Polyamide
Production time: N/A
Budget: $2,750
Edition: N/A

→ #building blocks

Creating Designs from the Smallest Intersection

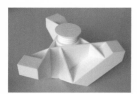

Keystone joint

The Rotterdam studio Minale-Maeda is deeply invested in the curiosities and potential of multi-directional material translations, from digital to analog and from complete object to unitary system, as well as the dual nature of objects as unique pieces and schematic blueprints for reproduction. Building on a download distribution platform explored in a previous project, Inside-Out Furniture, Minale-Maeda designed Keystones as a reduction of the consumer object to the bare essentials necessary for their function. The various Keystones bring together pieces of wood (and other materials) into precise configurations, including a table, a chair, and a coat hanger, using screws to fix the different members in place through tension, thus also creating some tolerance for variation in thickness. As a result, only the connector itself needs to be sold as a proprietary product: the other structural members can be sourced locally, cut to any length the user desires, and assembled without the need for joinery skills. Even the connector itself is a product of distributed design: rather than create them in a central location and ship them to customers, Minale-Maeda can send the digital files to a local printing service, where they can be manufactured and shipped directly to the user.

Michael Bernard
Nooks

→ #building blocks

Simplifying Furniture for Tolerant Assembly

Year: 2013
Client/Purpose: student research
Partners: University of Fine Arts of Hamburg
Technique: Fused deposition modeling
Printer/Service: RepRap Mendel
Material: PLA bioplastic
Production time: 1 year
Budget: N/A
Edition: N/A

Commercial furniture is often consciously designed with proprietary connections or structural systems in order to monopolize the spending potential of the customer: if a product cannot be combined with items from other companies or easily modified with standard elements, then it naturally encourages people to buy entire sets from one manufacturer (consider the entire platform of tools and components included in IKEA kits or the special screws used by electronics companies to prevent cases from being opened). While completing his master's thesis at HFBK Hamburg, German designer Michael Bernard wanted to explore a contrary autonomy of furniture elements that could work flexibly in tandem. Among the series of "half-products" he designed, Nooks is one of the simplest: consisting only of trapezoidal brackets, the system allows boards of any length, width, and thickness to be assembled as a variable structure that can be expanded or recombined at will without the need for any tools – not even a screwdriver. By virtue of the process of additive manufacturing, the brackets can be quickly modified to fit any material at no additional production cost. Bernard's project repositions users as designers with the potential to tolerate constant change and adaptation to unique circumstances.

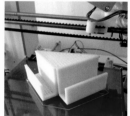

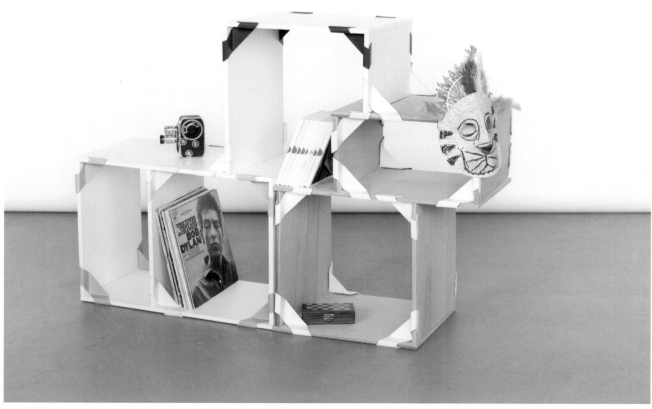

The Nooks system can be completed with any flat material and constantly adapted or expanded

Patrick Jouin
SOLID collection

→ #complexity

Innovative Structures for Single-Print Furniture

As with all transformative technologies, each advancement in the potential of 3D printing – from the speed or diversity of material to the resolution of the final finish – triggers an in-kind reaction. Thus, the increasing scales at which stereolithography and selective laser sintering could be

the customary design phase of adapting products for mass manufacturing: in this case, the prototype is also the final result.

practiced have generated an associated interest in full-scale furniture fabrication printed in a single pass. Considering that furniture design was, until recently, inseparable as a discipline from the craft of joinery, this innovation has precipitated enormous implications for industrial design. In 2004, French designer Patrick Jouin worked with Materialise to realize the SOLID collection of single-print, full-scale furniture, each piece exploring a different structural technique made possible by additive manufacturing. The S1 stool, for example, is a hollow cube with a surface perforated by bubble-like holes and internal bone-like members for support, while the C1 chair resembles a distorted honeycomb, vertically extruded to form a seat and backrest. The T1 table and C2 chair, on the other hand, use a language of apparently random ribbons, creating an overlapping network that is nevertheless capable of withstanding the body's weight. Through 3D printing, Jouin has avoided

Patrick Jouin
One_Shot.MGX

→ #complexity   → #assembly free

Folding Design to its Smallest Dimensions

The capacity of 3D printing to produce articulated objects in one pass may be considered one of its greatest advantages over other methods of production, negating the tedious and error-prone processes of automated and hand assembly. A less obvious outcome of this feature is the potential to print objects in their folded and compressed forms, thus enabling more units to be printed at once, reducing costs, energy consumption, and in some instances waste material (although most additive manufacturing processes have negligible waste). French designer Patrick Jouin created One_Shot. MGX in collaboration with printing company Materialise as a paradigm of elegant mechanical design solutions. The stool weighs just three kilograms and has a set of single-hinged legs printed in laser-sintered polyamide. One_Shot.MGX uses no axle, screws, or springs: it simply unfolds into a position fixed by gravity with delicate articulations hidden at the center. By printing the stools in the folded position, 15 units can be manufactured simultaneously whereas only two unfolded stools could fit because the diameter of the open seat is three times larger. While the innovation is rather well known today, it was a revolutionary design in 2006, acquired by the most significant design museums around the world, including the V&A in London, the Cooper-Hewitt in New York, and the MoMA in New York.

Patrick Jouin
Bloom

→ #assembly free   → #complexity

Floral Rhythms in Flexible Lighting

Year: 2010
Client/Purpose: For sale
Partners: .MGX by Materialise
Technique: Selective laser sintering
Printer/Service: Materialise
Material: Polyamide
Production time: N/A
Budget: $2,000 per print
Edition: 99

The relationship between forms found in nature and lighting design has a fruitful history: when copied, the branching or blossoming structures produce by some plants as a result of evolution also provide optimal light diffusion. This link, which has inspired such iconic examples as the 1958 Artichoke Pendant produced by Poul Henningsen for Louis Poulsen, was particularly suited to 3D printing given the breadth of formal complexity made possible by computational design approaches. Similarly, Janne Kyttanen's 2003 Lily.MGX lamp used the structural metaphor of flower blossom, becoming one of the most recognized examples of early 3D printing for consumer products. Patrick Jouin's Bloom table lamp, designed in 2010, joins this lineage of light fixtures, but it introduces the element of flexibility found in the daily cycles of flowers from morning to night. Each "petal" of the shade is hinged at the base, producing a nuanced spectrum of light, from direct to diffuse, that can be adapted by the user. In this work, Jouin, who explored a similar concept of functional mobility with his One_Shot.MGX folding stool in 2006, injects an almost traditional aesthetic motif with an element of user customization.

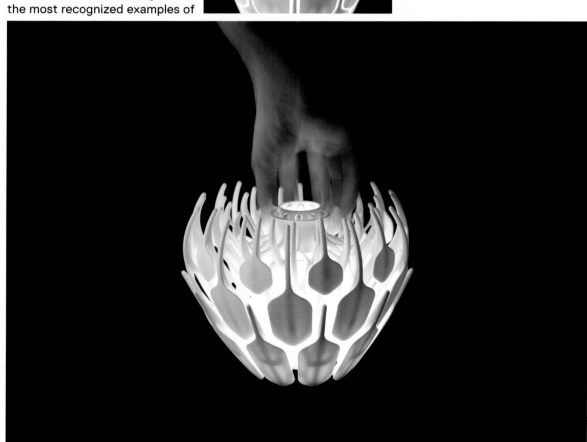

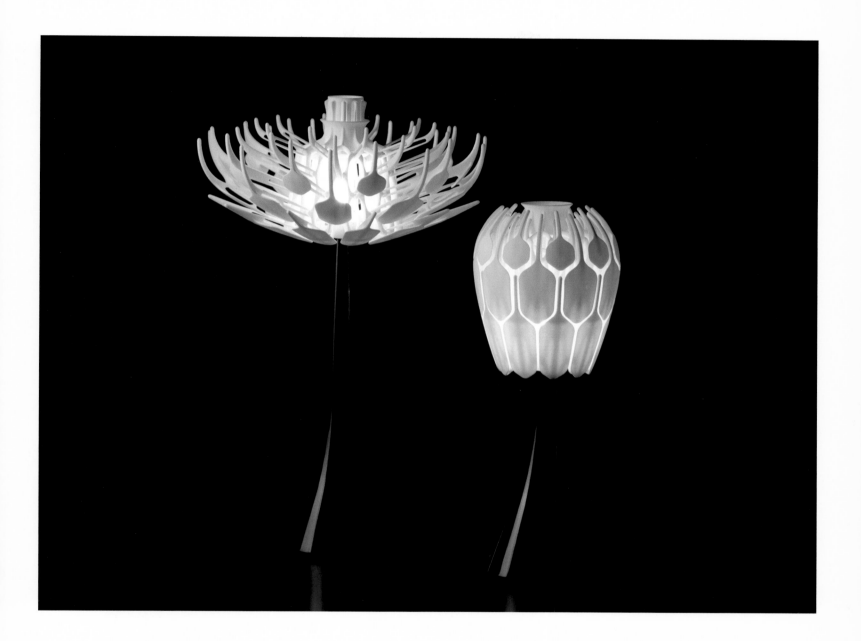

Stilnest
The Cuckoo Project

→ #complexity

An Endangered Craft Reimagined through Distributed Design

Year: 2013
Client/Purpose: Research
Partners: Daniel Hilldrup, Michiel Cornelissen, Mendel Heit, Arte Joyas, StudioSinth, Cipres Technology Systems, Phil Eichinger
Technique: Selective laser sintering
Printer/Service: Hasenauer & Hesser
Material: Polyamide
Production time: 5 weeks
Budget: N/A
Edition: 1

The cuckoo clock is a curious object. It is at the same time an artifact of the apotheosis of human mechanical engineering and an endangered species that is extremely vulnerable to technological changes. This tension is especially visible in the late twentieth century innovation of quartz watches, which operate electromechanically, using a battery rather than a spring – a development also known as the Quartz Revolution or the Quartz Crisis. The archetypal cuckoo clock established in the nineteenth century is now protected by the Black Forest Clock Association. Meanwhile, the German design company Stilnest has begun to reexamine the object in a contemporary context, using the technological facilities of networked communication, virtual modeling, and rapid prototyping to achieve an unprecedented outcome. The Cuckoo Project draws together the skills of six different designers from London, Utrecht, Berlin, Mexico City, Ghent, and Düsseldorf to design each element of the cuckoo clock – the pendulum, horn, house, rooftop, and bird itself, as well as the sound – in effortless synchrony. Given the ubiquity of clock and time-keeping functions on everyday objects today, Stilnest's cuckoo clock takes up the more whimsical task of striking on tweets,

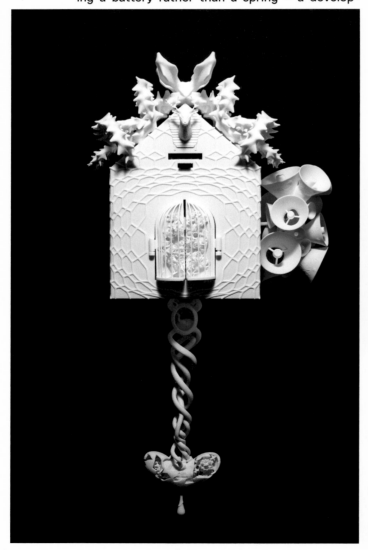

Elements of the Cuckoo Project are designed by people in different locations

Josh Harker

Crania Anatomica

→ #new industries   → #complexity

Crowdsourcing Sculpture through On-Demand Prints

Year: 2011
Client/Purpose: Artwork
Partners: N/A
Technique: Selective laser sintering
Printer/Service: Shapeways
Material: Polyamide
Production time: 6 months
Budget: $80 to $600 per print
Edition: On-demand and limited edition

When 3D printing is used as a technique for art, what are the implications for the artwork as a unique object? The cultural evolution related to abstract art (which minimized and even challenged the importance of the artist's signature on the work) and performance art (which reduced the focus on art as a concrete object and translated it into a script that could be performed anew endlessly) have altered the popular conception of what creative activity means. At the same time, the ambiguity of where the artwork lies – in the virtual model, in the first print, or in every print? – poses additional questions regarding a definition of the work as art. Sculptor Josh Harker is one example of a new kind of practitioner in the field who relies not only on rapid prototyping but also on networks to sustain his work. In 2011 he

wanted to create a piece that would push the limits of what could be achieved with consumer 3D printing: he created Crania Anatomica, a delicately filigreed skull, and took to crowdfunding platform Kickstarter to raise money for its realization, declaring that he would make as many skulls as were ordered in the 45 days of the campaign before submitting the piece to any gallery or exhibition. The project was launched with an initial goal of $500, but it ended up earning $77,271 in pledges (including 894 pre-orders for skulls), eventually becoming the highest grossing sculpture campaign ever made on Kickstarter.

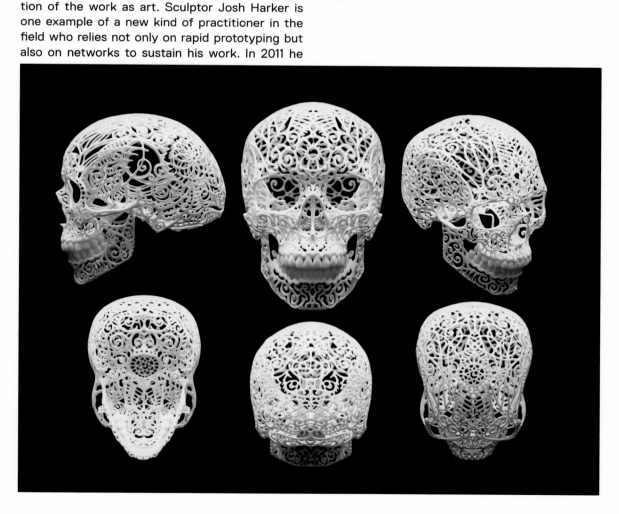

Nendo
Lacquered Paper-Objects

Year: 2012
Client/Purpose: For exhibition
Partners: Nilufar Gallery
Technique: Laminated object manufacturing
Printer/Service: N/A
Material: Paper, glue, lacquer
Production time: N/A
Budget: N/A
Edition: N/A

→ #materials → #new craftmanship

The Subtle Grain of Reassembled Paper Wood

A great deal of 3D printing is performed using white ABS plastic, a material that is almost without quality, lacking any inherent texture or color. Unfortunately, this tends to shape the our general understanding of 3D printing as a process that might be able to produce complex forms, but not objects that we feel any particular attachment to, as we would with traditionally crafted pieces. Yet Nendo, the inventive Japanese design studio, has proven the opposite with Lacquered Paper-Objects, a series of small vessels featuring an intriguing wood-like grain that suggests a tech-

nique refined over hundreds of years rather than the innovative application of CNC technology. In fact, the objects are made using a 3D printer that cuts, layers, and glues the form of a vessel within a stack of white paper. The object is then "carved out" from the leftover paper and sanded by hand. Finally, a black lacquer is painted onto the surface of the objects, leaving a layered texture that resembles finely sculpted wood. Nendo's approach to 3D printing shows that, for the thoughtful designer, neither the automation of processes nor the standardization of input materials should constitute a substantial barrier to the injection of mystery and wonder into the human encounter with the final object.

The glued paper layers are carved out from within the excess material

When sanded and lacquered, the laminated paper resembles wood

Ilona Huvenaars
Knitted Vase

→ #complexity   → #assembly free

Flexible Structures from Rigid Prints

Year: 2009
Client/Purpose: Research
Partners: Willem Derks
Technique: Selective laser sintering
Printer/Service: N/A
Material: Nylon
Production time: 7 months
Budget: $5,200
Edition: 1

Although 3D printing is widely recognized for its ability to produce articulated objects, it is not readily identified with elastic or flexible structures: today, the most advanced 3D printers can use rubber-like materials, but it is still a high-end option. In 2009 Ilona Huvenaars set out to make a design that could adapt in size, stretching like a knitted fabric despite the rigidity of the laser-sintered nylon material that was generally available at the time. Her Knitted Vase, designed in collaboration with Willem Derks and inspired by the interlocking loops of yarn found in knitwear, is printed in a single piece. The solid bottom provides a stable base, while the open pattern of entwined, printed strands at the neck of the vase can compress or expand to hold one flower or a whole bunch. Blending Huvenaaars's interest in traditional crafts and patterns with Derks's proficiency in 3D modeling and printing techniques, the Knitted Vase explores a formal approach that is only possible with the medium of rapid prototyping.

Tjep.

Signature Vase

→ #customization

Year: 2003
Client/Purpose: For sale
Partners: Droog
Technique: Selective laser sintering
Printer/Service: N/A
Material: Nylon
Production time: N/A
Budget: N/A
Edition: N/A

Personal Handwriting Translated into Physical Volumes

When Dutch design became a formidable international presence in the late 1990s, it challenged a prevailing context for domestic furniture that prioritized the diffusion of declarative design objects such as Philippe Starck's Juicy Salif lemon squeezer (still one of the iconic representations of design in the popular consciousness), at the height of efficiency in mass production. The alternative approach emerging from the Netherlands blended witty expressions of concept, heterogeneity, and the charming details and flaws characteristic of handicraft. 3D printing was not yet a

Droog by Tjep., the Amsterdam studio founded by Frank Tjepkema. Here, the idiosyncrasies of the object were defined by the unique style of the customer's handwriting rather than the designer's technique. In 2003 the Signature Vase was still expensive enough to remain a prototype. A decade later, it would be second nature for someone familiar with 3D modeling to vectorize a signature, extrude the shape into a solid form, and upload it to a 3D printing service for home delivery. Recently, Tjep. even published a successful attempt to print a Signature Vase using their own low-cost

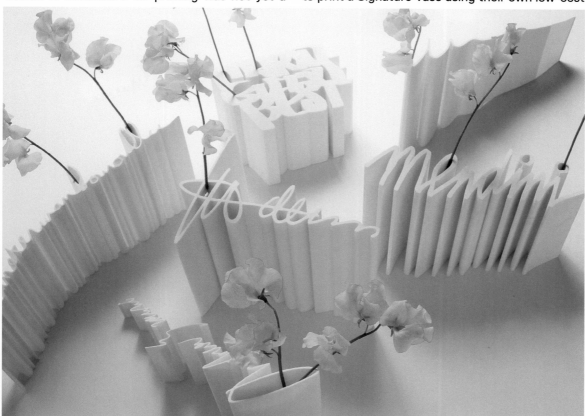

comfortable language for this approach, relegated more to the fields of experimental architectural models and innovative industrial prototyping. One of the earliest points of intersection between the two, however, was the Signature Vase made for

Ultimaker. The original, however, remains prescient for its time.

Dirk Vander Kooij
Endless Chair

Year: 2010
Client/Purpose: Student research
Partners: Design Academy Eindhoven
Technique: Fused deposition modeling
Printer/Service: Modified Fanuc robot
Material: Reused plastic
Production time: 3 hours per piece
Budget: $1,300 per print
Edition: On-demand

→ #new industries   → #new craftmanship   → #new industries   → #machines

The Instantaneous Beauty of Low Resolutions

Between the digital model and the printed object lies a matter of significant translation. The infinitely smooth planes of the virtual sphere are rendered in strata of fixed height; any reduction in the thickness of the printed line implies an increase in printing time. Very fine printing is therefore unsuitable for the mass production of furniture such as tables and chairs given the duration of large-scale printing. For his graduation project from Design Academy Eindhoven, Dirk Vander Kooij approached this issue from the opposite side: if low resolution were embraced as a virtue rather than a drawback, what potential could it offer? After acquiring a decommissioned Chinese robot, he decided to recreate a 3D printer on a titanic scale, using a thick deposition line (made of melted refrigerator plastic) to trace a full-sized chair. The Endless furniture series, which includes tables, chairs, and a lamp, is both robust and efficiently made. An entire chair can be printed in under three hours with a striated surface that reveals the course of the printing process. Unlike many experimental designers, Vander Kooij's creations are both prototypes and viable commercial products at the same time. Each print can be adjusted in terms of scale, profile, and function with negligible costs in terms of time and money. Even the minuscule amount of waste material generated by the robot is translated into a product: when changing input material, it prints clothes hangers that fade from one color to the next.

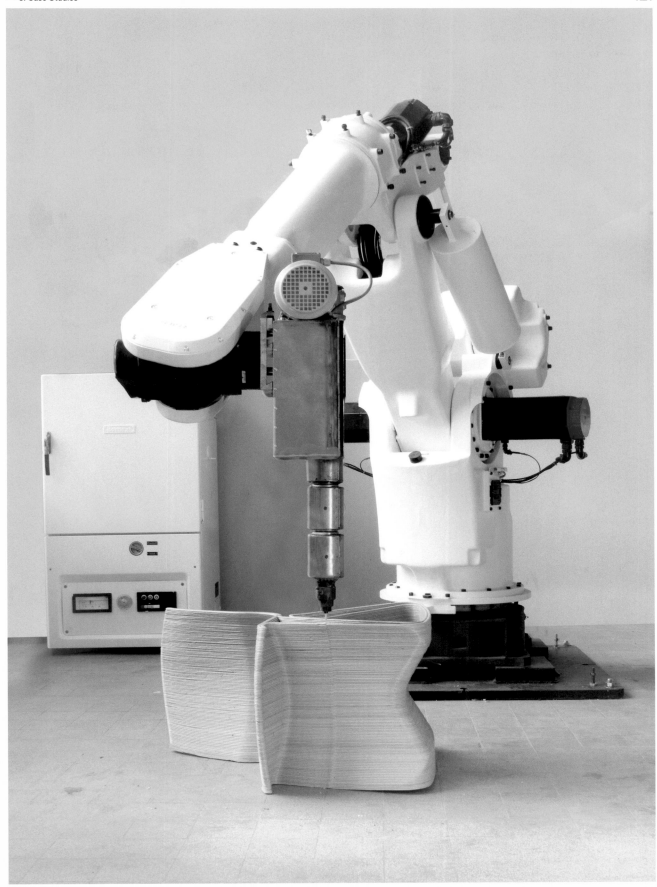

Each time an Endless chair is printed, the form can be subtly changed at no additional cost

WertelOberfell
Fractal.MGX

Year: 2009
Client/Purpose: For sale
Partners: .MGX by Materialise,
Matthias Bär
Technique: Stereolithography
Printer/Service: Materialise
Mammoth SLA
Material: Resin, UV-stable
pigment
Production time: N/A
Budget: $11,000
Edition: 20

→ #complexity

Growing Complexity into Full-Scale Furniture

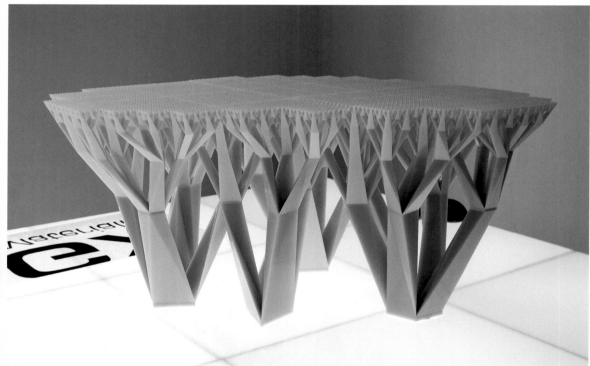

The complexity theories that guided advanced scientific study in the second half of the twentieth century began to exhibit visual manifestations in cultural fields of production, such as architecture and design, by the 1990s. As scripting overtook more traditional forms of composition among creative practitioners, the resulting buildings and furniture evoked the self-organizing structures found in the natural world more than the clean perpendicular lines of the modern era. However, this approach remained very difficult to realize using conventional industrial techniques, which favored repetition and standard modularization. Now that 3D printing can be used on a larger scale, allowing as much complexity as desired, designs that were once mere theoretical proposals on paper can now take concrete form. The German studio WertelOberfell took advantage of this new opportunity to study fractals found in the world of mathematics as well as nature. The first Fractal Table, made in 2008, adopted a branching language reminiscent of trees, with large dispersed roots culminating in a regular grid of points. A second iteration, Fractal.MGX, employs a denser upper surface to function better as a coffee table.

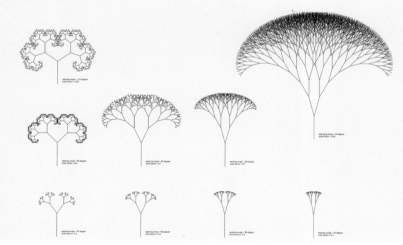

Complex growth patterns found in nature

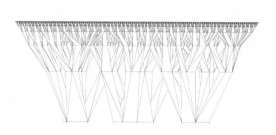

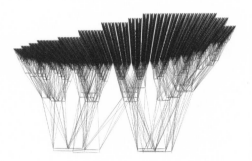

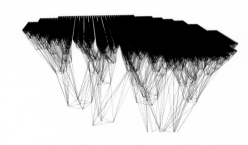

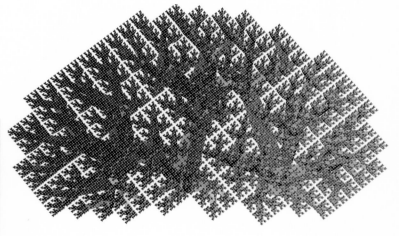

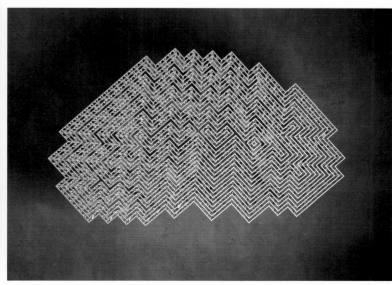

KRAM/WEISSHAAR
Multithread

Year: 2012
Client/Purpose: For exhibition
Partners: Nilufar Gallery
Technique: Selective laser sintering
Printer/Service: N/A
Material: 3D printed aluminium joints, high tensile strength steel tube, powder coated aluminium shelf surface
Production time: N/A
Budget: N/A
Edition: N/A

→ #building blocks → #software

Visualizing Forces in Optimized Structures

For all but the most advanced craftsmen, the difficulty of visualizing unseen forces during the making process encourages recourse to familiar and reliably stable structural forms, which in turn reinforces the fundamental furniture archetypes that human society has used for millennia. However, as parametric engineering software has diffused from the recesses of advanced industry (like aviation) to a much wider audience of architects and designers, it has also enabled a kind of informed improvisation within the virtual modeling environment. The design studio KRAM/WEISSHAAR has developed a "force-driven structure" methodology that emanates from this technology, using custom-built software to upend the conventional design approach. Multithread, for example, is a series of furniture that revolves around two elements – horizontal surfaces and structural webs. By de-

fining a particular placement of the horizontal element in space (such as a tabletop, shelf, desk, or other function), the designers can proceed to build the support for this surface, using real-time information on the forces acting within the object to carefully calibrate the performance of each joint and member. The ensuing manufacture is a similar hybrid of high-tech automation and fine operations performed by hand: the customized joints and tubes are 3D-printed in metal and then assembled by master craftsmen, who finally paint the object in colors that represent the stresses and strains at each point on the structure.

The metal joints are painted to indicate different degrees of structural force

Michiel Cornelissen
36 Pencil Bowl

Year: 2010
Client/Purpose: For sale
Partners: N/A
Technique: Selective laser sintering
Printer/Service: Shapeways, Materialise
Material: Polyamide
Production time: 3 months
Budget: $100 per print
Edition: On-demand

→ #building blocks

Connecting Standardized Objects with Distributed Printing

The role of the designer used to require proving the financial viability of a prototype to a mass manufacturer and making substantial modifications in order to meet the conditions of efficiency defined by the machines, materials, and shipping and distribution models available at the time and

services like Shapeways and Materialise (additionally, with no exclusivity on any particular printing platform). 36 Pencil Bowl, for example, is a simple connector piece that users can fill with standard hexagonal pencils: the design is based on the profile of pencils made by companies like Stabilio,

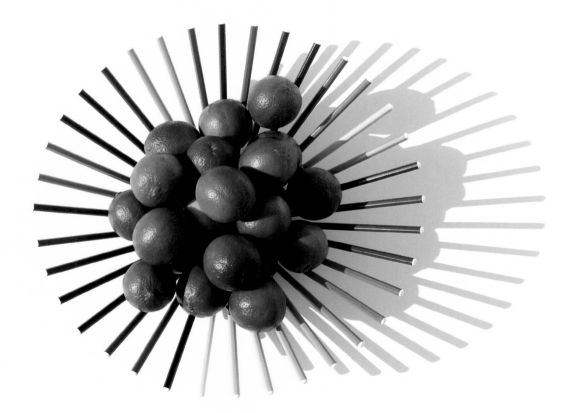

place of production. Today, designers have a very different set of apparatus at their disposal, including cheap web sites, a vast online marketplace, distributed methods of manufacture, and, last but not least, rapid prototyping. A designer like Michiel Cornelissen is an instructive case study: in addition to developing products for large companies like Philips and Bosch and boutique labels like Soonsalon, he offers his own series of playful design sketches for on-demand printing through

Caran d'Ache, and Prismacolor. When the circular fan of pencils is complete, the entire set can serve as a bowl for food or small household objects. Products like 36 Pencil Bowl can be released to the public immediately as soon as the design has been finalized. The designer incurs little risk but is able to experiment casually with different formal and logistical models of design.

Marco Mahler
3D-Printed Mobiles

→ #complexity → #assembly free

Kinetic Structures, Virtually Balanced

Year: 2013
Client/Purpose: For sale
Partners: Henry Segerman
Technique: Selective laser sintering
Printer/Service: Shapeways
Material: Nylon
Production time: 3 months
Budget: $3 to $1,103 per print
Edition: On-demand

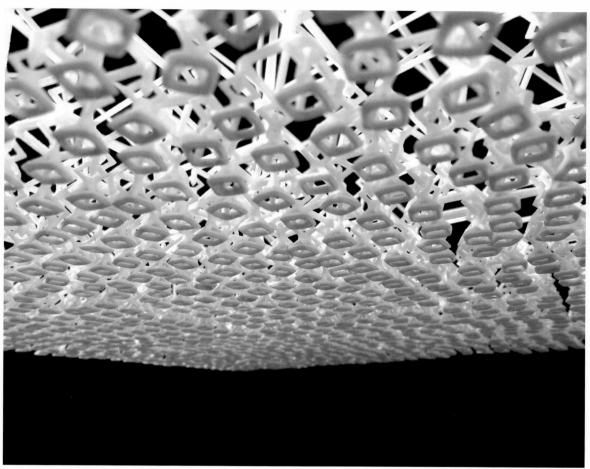

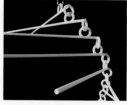

The term "mobile" was first used in 1931 by Marcel Duchamp to describe the kinetic wireframes made by Alexander Calder, whose iconic works later sparked an entire genre of sculpture. These works were created in situ, taking advantage of ambient air currents and finely calibrated equilibriums to generate large constructions, both dynamic and serene in their movements. In the years that followed, the art form became archetypal, but the advent of sophisticated design and engineering software has allowed for unprecedented methods of working in the medium. A collaboration arose between kinetic sculptor Marco Mahler and Henry Segerman, an assistant professor of mathematics, on the question of advanced modeling for the creation of rapid prototyped mobiles. Communicating exclusively via Twitter and Skype, the two were united by a shared interest in a mobile emerging fully assembled from a single print, the separate parts already interconnected. The precision of the printing technology allowed them to calibrate the balance to the micron (one one-thousandth of a millimeter), slightly modifying the thickness of certain elements to generate a different bearing. They also experimented with scripted modeling in the Python programming language to increase the limits of intricacy, for instance with the 1,365 pieces that make up the Quaternary Tree Mobile Level 5. Finally, they uploaded their designs to Shapeways, transforming the mobile from a delicate one-off to a geometric code available for materialization on demand.

Front
Sketch Furniture

→ #interact → #intangible

Wireframe Objects Drawn in Thin Air

Year: 2005
Client/Purpose: For exhibition
Partners: Playground Squad, Acron, Friedman Benda Gallery
Technique: Stereolithography
Printer/Service: N/A
Material: Photopolymer
Production time: 1 day per piece
Budget: N/A
Edition: N/A

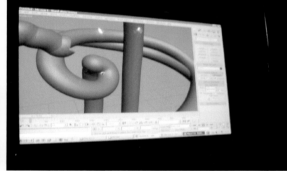

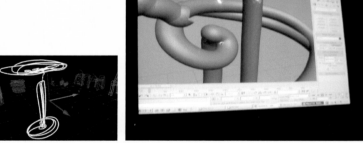

The term "rapid prototyping" obscures the fact that the transition between ideas and 3D printing still requires the ability to "speak" the language of 3D modeling software, a skill that is mostly the reserve of designers, architects, and engineers. This enforces certain biases, for instance symmetry and smoothness, which have more to do with the limitations of the virtual environment than the predispositions of the physical world. Front, a Swedish design collective, decided to forgo conventional 3D modeling programs in favor of motion capture technology when designing their Sketch Furniture series: by drawing "blindly" in real space with an electronic pen, the designers were able to generate 3D drawings of a chair, table, and lamp. These files were then printed with no digital modification: a laser traced the gestural motions in liquid plastic, leaving a hardened outline of the furniture sketches. The final results, coated in white lacquer, reveal the largely unexploited potential of 3D printing to quickly and faithfully materialize the initial "spark" of an idea when filtered through a precisely adapted, if unconventional, medium.

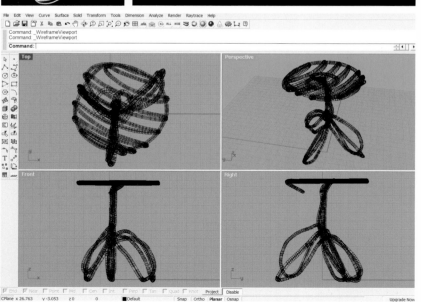

The motions traced by hand are captured on the computer as geometric paths

Fung Kwok Pan
Fluid Vase

→ #customization

Materializing the Flow of Water as a Printed Vessel

Year: 2010
Client/Purpose: Student research
Partners: Nanyang Technological University, Fabrizio Galli, Chua Chong Han
Technique: Selective laser sintering
Printer/Service: N/A
Material: Polyamide
Production time: 9 months
Budget: N/A
Edition: N/A

The rise of Dutch design in the late 1990s and early 2000s saw all manner of objects, from banal plastic cups to pistols, being cast in plaster and remade in blank white ceramic in an attempt to challenge the meaning of things through material transformation. Fung Kwok Pan took on the arguably much more difficult task of reproducing a splashing liquid, thus creating a vase from the outline of water being poured into it. Of course, being unable to freeze the volatile material in a single instant, the Singapore-based designer turned to the mathematical simulation of liquid motion in 3D modeling software, a technique that had been made increasingly precise in tandem with the rise of CGI in film and game production. Rather than simply choosing an arbitrary form and manufacturing it en masse, however, Pan developed an online interface that allowed people to isolate a unique form using three parameters: the shape of the enclosing container (square, circular, or triangular), the height from which the water is poured, and the desired moment during the pouring process. The resulting virtual model could then be 3D-printed in laser-sintered polyamide. The Fluid Vase is thus an object with no predetermined form, but an algorithmic code that is recalculated every time an item is made.

1. Select a container shape.

2. Determine position of pour.

① **②** ③

3. Determine volume of flow.

4. Choose your favourite frame.

33

PREVIEW

The parameters for the Fluid Vase can be customized by each user

1. Select a container shape.

2. Determine position of pour.

① **②** ③

3. Determine volume of flow.

4. Choose your favourite frame.

33

x €399.00

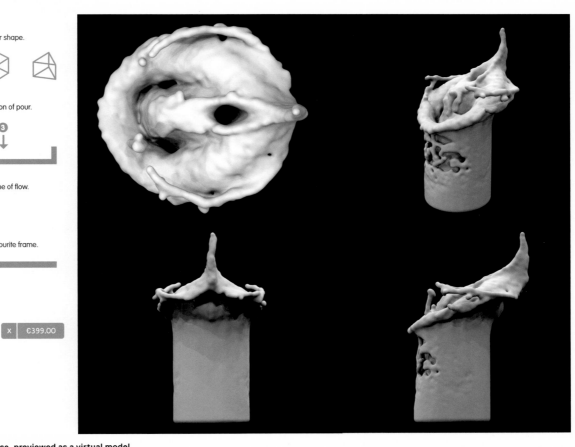

The customized vase, previewed as a virtual model

FormNation/Jan Habraken
Chairgenics

→ #software

Breeding the Perfect Chair through Parametric Modeling

Year: 2013
Client/Purpose: Research
Partners: Uformia, Mathieu Sanchez
Technique: Stereolithography
Printer/Service: N/A
Material: Resin, polyurethane foam
Production time: 5 years
Budget: $15,000
Edition: N/A

The replicable nature of 3D printing makes it inherently well-suited to experiments in iterative design: as a product goes through successive versions, models can be automatically produced to chart the evolution in form. Jan Habraken, the founder of FormNation, has exploited this capacity to work on chairs in an unconventional way. Rather than approach the design of a chair from an authorial perspective, his team established Chairgenics as a platform for empirical research into the "gene pool" of chairs already in existence. By assigning numerical values to each chair based on comfort, durability, design, cost, and aesthetics (measured by the number of online search results), FormNation could hypothesize an ongoing evolution for the archetypal chair, similar to the process that takes place in living species. In collaboration with Uformia and Mathieu Sanchez, this optimization process was automated in a virtual 3D modeling environment. Over time, the team was also able to refine the process to more accurately mimic the Darwinian principle of survival of the fittest, including the preservation of successful traits and the incompatibility of certain partners. Finally, some examples of the chair were realized in epoxy resin using stereolithography, later filled with polyurethane foam for strength. Despite this progress, FormNation's quest to breed the perfect chair continues.

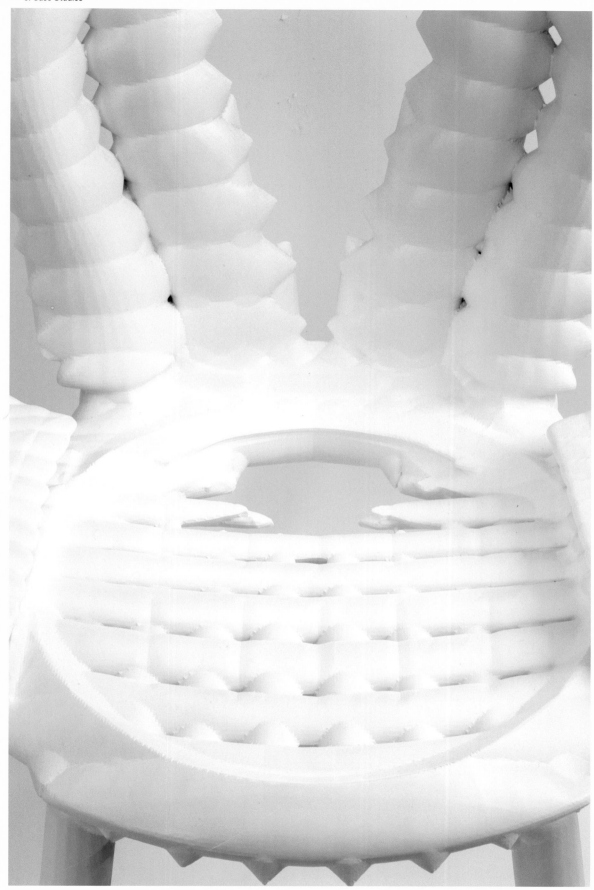

Chairgenics breeds different chairs to evolve new descendants

100% archetypical × 100% Bertoia

100% archetypical × 100% Tulip

90% archetypical
10% Bertoia

10% archetypical
90% Bertoia

× 100% Monoblock

60% archetypical
40% Tulip

40% archetypical
60% Tulip

× 100% Tom Vac

3% archetypical
27% Bertoia
70% Monoblock

9% archetypical
81% Bertoia
10% Monoblock

× 100% Eames Hardshell

100% Chair One ×

36% archetypical
54% Tulip
10% Tom Vac

20% archetypical
39% Tulip
50% Tom Vac

× 100% Plia

100% Thonet ×

3% archetypical
25% Bertoia
3% Monoblock
70% Eames Hardshell

9% archetypical
81% Bertoia
10% Monoblock
30% Eames Hardshell

19% archetypical
26% Tulip
5% Tom Vac
50% Chair One

10% archetypical
15% Tulip
25% Tom Vac
50% Chair One

5% archetypical
40% Bertoia
5% Monoblock
10% Eames Hardshell
40% Thonet

Genealogy tree

100% archetypical   ×   100% Pantone

100% archetypical   ×   100% Tolix

100% archetypical   ×   100% Eames Hardshell

100% Zig Zag   ×   10% archetypical   90% archetypical
               90% Pantone   10% Pantone

70% archetypical   50% archetypical   ×   100% Pantone
30% Tolix   50% Tolix

100% Thonet   ×   30% archetypical
                 70% Eames Hardshell

†   1% archetypical
    9% Pantone
    90% Zig Zag

100% Seven   ×   25% archetypical
                 25% Tolix
                 50% Pantone

21% archetypical
49% Eames Hardshell
30% Thonet

12% archetypical
50% Seven
25% Tolix
13% Pantone

Estudio Guto Requena
Nóize Chairs/Cadeiras Nóize

→ #intangible

Real-Life Sounds Disrupt the Icons of Modern Furniture

Year: 2012
Client/Purpose: For exhibition
Partners: Coletivo Amor de Madre, META-D
Technique: Fused deposition modeling
Printer/Service: N/A
Material: ABS plastic
Production time: 3 months
Budget: N/A
Edition: 3

The availability and free exchange of data made possible by peer-to-peer networks and the increasing digitization of content have caused paradigm shifts in many formats, from MP3s to 3D files. With Nóize Chairs, Estudio Guto Requena in São Paulo charts the intersection between music and form as fertile ground for copying, sampling, and remixing. The project began by modeling three iconic Brazilian chair designs in 3D: the Girafa chair by Lina Bo Bardi, Marcelo Ferraz, and Marcelo Suzuki, the Oscar chair by Sério Rodrigues, and the São Paulo chair by Carlos Motta. Later, the studio recorded sounds at three locations in the city of São Paulo (Grajaú, Cidade Tiradentes, and Santa Ifigênia) and then used the audio files to distort the virtual chair models with a script written in Processing. Finally, the digital files were printed at full scale in ABS plastic, fusing the input of a distributed network of designers, modelers, engineers, and passersby on the street. A further development of the project, Live Nóize, demonstrated the variability of possible outcomes by restaging the design process in the Coletivo Amor de Madre Gallery. Visitors could contribute their own sounds and witness the distortion of the virtual model in real time. Each day, a miniature Girafa chair was printed to create a hybrid archive of sound-sensitive design.

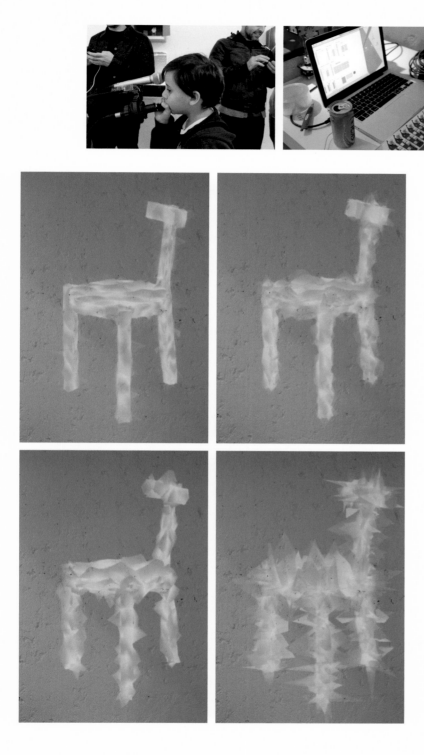

Live Nóize: A Design Performance, **São Paulo Design Weekend 2013**

Amanda Ghassaei
3D-Printed Record

Year: 2012
Client/Purpose: Research
Partners: Instructables
Technique: Polyjet
Printer/Service: Objet500 Connex
Material: VeroWhitePlus acrylic photopolymer
Production time: 3 months
Budget: $100 per print
Edition: In development

→ #software   → #open source

Pushing Precision and Data Density to a Microscopic Scale

A Processing script converts audio data into geometric grooves

3D printing is ideal for creating complex forms that would be too laborious to make using traditional manufacturing methods. Whereas some projects simply exploit this potential for sculptural complexity, others delve into the possibility of encoding vast amounts of functional data into a physical object. A clear example of the latter is the 3D Printed Record by Amanda Ghassaei. Using a high-resolution printer to create records in resin cured with ultraviolet light, Ghassaei was able to test the limits of additive manufacturing against standard vinyl-cutting technology. However, the real innovation lies not in the printing process itself but in the virtual transformation of raw audio data into a digital 3D model using a custom script built in Processing and the ModelBuilder library designed by Marius Watz. Over three months, Ghassaei refined the algorithmic generation of a geometric groove pattern, taking into account the limitations of additive manufacturing in order to arrive at the best possible sound when the record was played on a turntable. The script has since been made available online as a free, open-source document so that others can experiment with their own 3D printers and unique audio samples.

Kathryn Hinton
Digital Hammer

→ #interact

Manipulating Virtual Models with Physical Force

While the first wave of 3D printing could be distinguished from more mechanical or manual techniques by the removal of the designer's physical and metaphorical fingerprints from the object, current trends are shifting the focus to more nuanced methods of design, taking advantage of the increasing sensitivity and range of both motion capture software and haptic tools for converting force into digital input for computers. In that vein, designer and silversmith Kathryn Hinton has been working for several years to forge an intersection between traditional metalcraft and rapid prototyping in The Craft of Digital Tooling, her graduation project from the Royal College of Art. Hinton was inspired to digitally emulate the silversmithing technique of raising, in which a sheet of silver is formed through hammering. By tracking the force and direction of the strikes made with the digital hammer, the computer models a semi-analog design process, displaying the virtual model in real time as an outcome of the recorded motion. When satisfied with the form, Hinton prints the model in wax and then uses the lost-wax method to cast sterling silver in the same shape. While the bowls, plates, and vessels made in this way are conventional archetypes for silversmithing, their faceted, angular aesthetic is unusual to the craft. In fact, they could not be made by hand.

The virtual model is printed in wax and then cast in silver

Year: 2008 – 2014
Client/Purpose: Research
Partners: Royal College of Art
Technique: Material jetting (wax)
Printer/Service: Solidscape
Material: Wax, sterling silver
Production time: N/A
Budget: N/A
Edition: In development

Unfold
Stratigraphic Manufactury

→ #new craftmanship   → #open source

The Unique Idiosyncrasies of Digital Production

Year: 2012
Client/Purpose: For exhibition, research
Partners: Jonathan Keep, Eran Gal-Or, Ahmet Gülkokan, Mustafa Canyurt, Eric Hollender, Larisa Daiga, Jen Poueymirou, Benjamin Matthews, Alicia Ongay-Perez
Technique: Paste extrusion (ceramic)
Printer/Service: Modified Bits from Bytes Rapman
Material: Ceramic, mixed media
Production time: N/A
Budget: $4,000
Edition: N/A

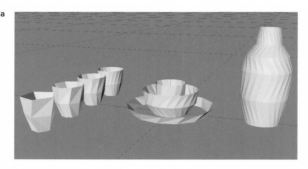

The premise of 3D printing as a universally identical technique is somewhat mythical, both in terms of the physical constraints of mechanically imperfect machines and the limitless opportunities for modifying open-source printers or experimenting with materials. While large-scale printing services like Shapeways and extremely precise processes such as photopolymer jetting are poised to offer "perfect" results, humble personal printers at the other end of the spectrum could better be described as a tool for cottage industry or craft. Unfold's Stratigraphic Manufactury attempts to capture the beauty of imprecision and experimentation in the translation of identical virtual models into individually manufactured objects. Working with a distributed global network of craftsmen,

Unfold has created local production centers revolving around their ceramic 3D printer in Istanbul, New York, London, Tel Aviv, and Suffolk. The ceramicists in each location are given an identical set of 3D files of cups, saucers, and vases, and are asked to print them without modifying the digital files. However, they are free to alter all other conditions, from the type of clay or porcelain used to the humidity, temperature, calibration, and layer thickness. Both intentional variations and fortuitous mistakes – glitches rendered in lacy tendrils of misaligned clay – are added to a growing set of international specimens, presenting the unexpected heterogeneity of 3D-printed ceramics.

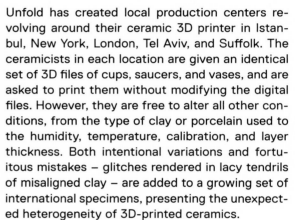

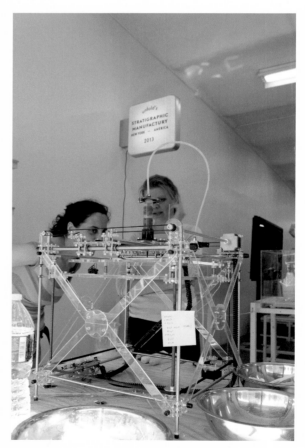

A series of identical virtual models is made by different designers around the world with local material variations

Jonathan Keep
Icebergs

→ #software → #new craftmanship

The Intricacy of Nature in 3D-Printed Clay

Year: 2013
Client/Purpose: Research
Partners: N/A
Technique: Paste extrusion (ceramic)
Printer/Service: Self-built Delta
Material: Porcelain
Production time: 2 to 3 days per piece
Budget: $750
Edition: N/A

Process

3D printing ceramics using a self-built Delta printer

The use of ceramics in additive manufacturing was, in some sense, inevitable given the mutability of the material, from soft and ductile to hard and brittle, depending on temperature and moisture. While designers approached the issue from a technological point of origin, British ceramicist Jonathan Keep was interested in the possibility of substantially increasing the complexity of his work as a craftsman. He began by using a handheld syringe as a clay extruder, making coiled pots that referred to the virtual design language of Boolean operations and lofted curves. Eventually, his research brought him into contact with Unfold, who had modified a RepRap 3D printer to deposit clay, and Keep was able to build upon his knowledge to fabricate his own printer. Using Java and Processing scripts, his work attempts to create natural patterns in morphological families of objects, all related by a digital genetic code. His Icebergs series, for example, echoes the apparently random form of its naturally occurring namesake, using a textural and formal intricacy that would be nearly impossible to produce by hand – although he also uses traditional manual techniques to overcome the limits of 3D printing, such as the size of the printable envelope. For Keep, this approach is not contrary to the traditional essence of ceramics, where the maker as authorial artist has always had to interact with the material itself, in all its physical and chemical complexity, working with natural affinities and overcoming pressures.

Icebergs are 3D-printed and assembled by hand for larger constructions

IN-FLEXIONS, François Brument, and Sonia Laugier
Vase#44

→ #interact   → #intangible

A Vase Formed in the Shape of a Voice

Year: 2009
Client/Purpose: For exhibition
Partners: N/A
Technique: Selective laser sintering
Printer/Service: N/A
Material: Polyamide
Production time: 2 months
Budget: N/A
Edition: On-demand

The essentially seamless flow from digital design to physical manufacture that characterizes 3D printing is also an invitation to designers to alter their way of working. For example, parameters rather than fixed values can be introduced early on to increase the variety of the final output. Depending on the interest of the designer, these parameters can be tied to any quantifiable variable, whether or not it is intrinsically related to the object of design. Ultimately, the approach can become more of a framework for interpreting different inputs than a linear progression of single-author creation. Francois Brument, the founder of IN-FLEXIONS, created Vase#44 as a way to directly harness the participation of the user in order to generate form. The studio built custom software to record and process the sound of each person's voice in order to shape a personalized vase. The height is determined by the length of the audio sample and the width by the sound volume, while the profile reflects the undulations of the sound frequency itself. Finally, the vase is printed in laser-sintered nylon as a record of the encounter between each user and the design apparatus.

Vase#44 installation

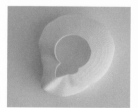

IN-FLEXIONS/François Brument and Sonia Laugier
KiLight

Year: 2011
Client/Purpose: Research
Partners: N/A
Technique: Selective laser sintering
Printer/Service: N/A
Material: Polyamide
Production time: 2 months
Budget: N/A
Edition: N/A

→ #interact

Customized Light Fixtures from Kinect Body Captures

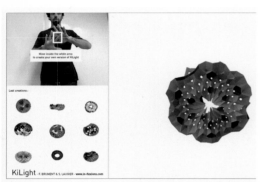

Interface

While 3D modeling is rapidly becoming a universal language given the growing ubiquity of interactive digital environments and games, the traditional techniques of craft remain reliant upon ingrained muscle memory and fine motor skills learned through experience and specialized training. However, the relative independence of the design and manufacture phases in rapid prototyping allows for a broader range of physical motions to indirectly generate form, even when performed by untrained "amateurs." After their experiments with modeling based on audio samples in the Vase#44 project, the French studio IN-FLEXIONS looked into more tangible interpretations of user input. Using Microsoft's Kinect, a motion-sensing device for hands-free game interaction, they devised an interface that would allow individuals to design their own KiLight. Users can create shapes, intuitively and in real time, by moving their body or hands in front of the Kinect, and the resulting image capture determines the shape, cellular size, structure, and color of the pendant lampshade. When the desired result is achieved, users simply hold their position for five seconds to save the 3D file. The unique KiLight can then be printed directly in laser-sintered nylon.

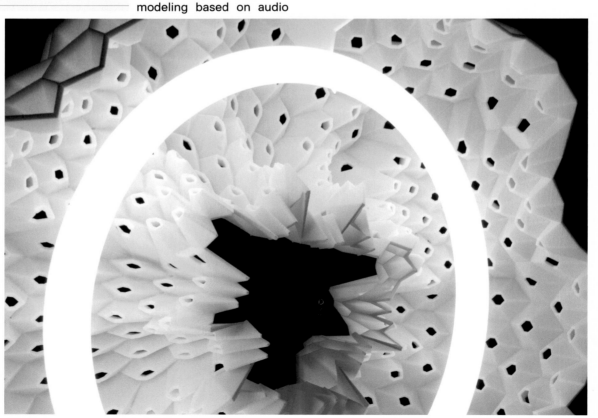

Detail

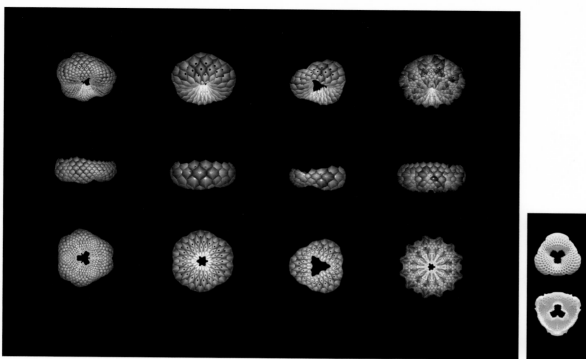

Variations

SLS parts

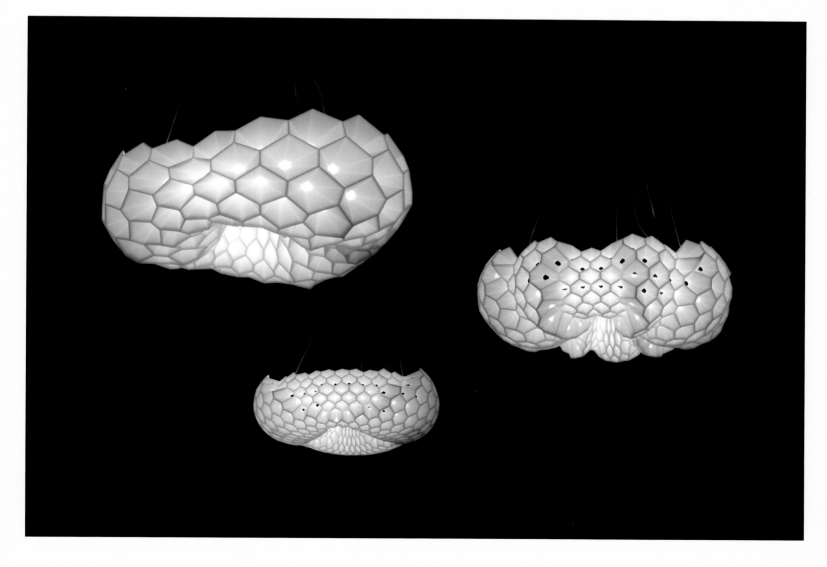

Studio Homunculus and Joong Han Lee
Haptic Intelligentsia

→ #interact   → #intangible

Formal Approximations in Hand-Printed Hot Glue

Year: 2011
Client/Purpose: Student research
Partners: Design Academy Eindhoven
Technique: Haptic force feedback deposition modeling
Printer/Service: Geomagic Touch Haptic Device, modified hot glue gun
Material: Hot melt adhesive polymer
Production time: 4 months
Budget: $7,000
Edition: N/A

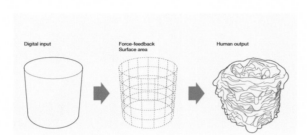

The texture of the hot melt glue

A printed object

While 3D printing may have an easier learning curve than a craft such as ceramics, it lacks a strong sense of feedback between technique and form-making: for amateurs, it can be hard to develop a sense of proportion and scale in the vacuum of digital environments. For his master's thesis at Design Academy Eindhoven, Joong Han Lee (the founder of Studio Homunculus) explored the idea of physical feedback between technology and the maker, not simply as a learning experience but as a method of production in itself. In Haptic Intelligentsia, a hot glue gun is attached to a haptic device that tracks the motion of the user's hands in space. When the tip of the device comes into contact with a predetermined virtual shape, it sends a vibration signal to the user, providing a rough sense of the shape and size of the desired object. The glue gun is used like a low-tech fused deposition modeling device, although the consistency and dimension of its flow are much more variable than the output of a 3D printer. The resulting shape, built from layers of solidified glue, is therefore a sensitive negotiation between a "perfect" digital blueprint and a more tactile analogic process, reflecting the accuracy and dexterity of each user's motions.

A user interacting with Haptic Intelligentsia

The user approximates a digital form in real space through haptic feedback

The same digital model printed by different users

Eric Klarenbeek

MyceliumChair

→ #materials → #sustainable

Growing Furniture with Organic Matter and Living Glue

Year: 2013 to 2014
Client/Purpose: Research
Partners: Beelden op de berg, Wageningen University, CNC Exotic Mushrooms
Technique: Fused deposition modeling
Printer/Service: N/A
Material: Organically grown substrate, mycelium, bioplastic
Production time: 1 year
Budget: N/A
Edition: N/A

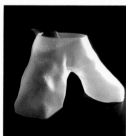

3D printing has the capacity to reduce material consumption in some design techniques (for example, by removing the need for molds and enabling on-demand production), but it still relies predominantly on petroleum-based plastics. In collaboration with Wageningen University's Mushroom Research Group and the Dutch company CNC Exotic Mushrooms, designer Eric Klarenbeek has begun experimenting with organic matter as the material input for additive manufacturing. His initial hypothesis on the possibility of printing living cells eventually led him to the design of a chair made of mycelium mixed with powdered straw and water, and coated with bioplastic to maintain the initial form. Over time, the water is replaced by the growth of mycelium – the mass of branching filaments that breaks down organic matter and sprouts mushrooms. Whereas conventional 3D printing involves melting plastic into a single mass, the printed substrate used for the MyceliumChair is fused together by the "living glue" of fungi. The colonization of living matter throughout the chair makes it strong enough to support the weight of a person. Klarenbeek's research unites 3D printing with a larger investigation into "growing" or "cultivating" the furniture and architecture in our environment, rather than simply fabricating it.

Several material tests

The organic substrate, made mostly of powdered straw

Mycelium grows over several days, acting as a binder

The mobile printing lab for mycelium-based objects

Nervous System
Colony Experiments

Year: 2013
Client/Purpose: Research
Partners: N/A
Technique: Inkjet powder printing
Printer/Service: Shapeways
Material: Colored gypsum
Production time: 3 weeks
Budget: N/A
Edition: 1

→ #complexity

Accentuating Form with Color

The geometric mesh, generated in Processing

The visual language of 3D printing was defined by single-material printing for such a substantial part of its early history that the recently increasing availability of full-color printing through commercial printing services has posed an unexpected challenge to designers. Most designers are accustomed to expressing themselves rather dexterously through form rather than hue, and even the interface of most 3D modeling programs encourages an abstract conception of colorless, transparent wireframes. Full-color printing seems somehow relegated to photorealistic manifestations, whereas its potential could be used for very different purposes. The Boston studio Nervous System created Colony as an investigation into how to exploit the possibilities of full-color printing and how to accentuate three-dimensional form through tone. Vaguely reminiscent of aquatic life-forms like coral reefs or anemones, the elaborately spined and lobed experimental objects are a skillful display of three-dimensional color gradients. Both the geometric mesh and the color were generated in tandem using Processing software. Colony hints at the beginning of a new language of surface complexity that could arise within the world of 3D printing.

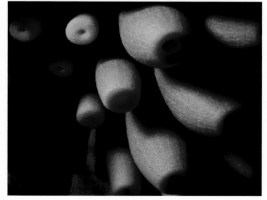

Olivier van Herpt
3D Printing Ceramics

Year: 2013
Client/Purpose: Student research
Partners: Design Academy Eindhoven
Technique: Paste extrusion (ceramic)
Printer/Service: Rostock Mini
Material: Earthenware, porcelain
Production time: 5 months
Budget: $2,000
Edition: N/A

→ #machines

Precious Imperfection as Aesthetic Artifacts

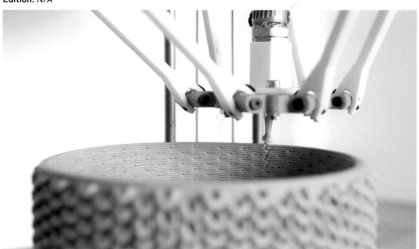

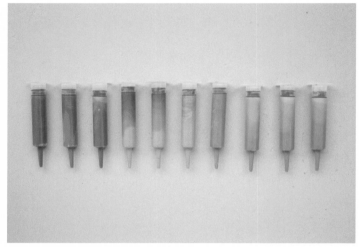

Syringes filled with colored clay

rapid prototyping to dripping beeswax and robotic drawing machines. Produced on a modified opensource Delta 3D printer, van Herpt's 3D-printed ceramics implement stochastic factors that give each piece its own unique form. At the same time, they also acquire a texture that is still rather foreign to the language of fused deposition modeling: by varying the thickness of the clay extruded by the machine, for example, van Herpt creates a surface with several grades of detail, similar to the three-dimensional bobbles in knitted fabrics, which despite the non-uniform appearance would be difficult (if not impossible) to make by hand.

Different types of clay can be used in the machine

Paradoxically, the daunting technical precision of 3D printing is also the factor that can make it the most monotonous, since the human eye tires of the perfectly identical patterns and shapes generated by machines. Meanwhile, the work of Olivier van Herpt, a student at Design Academy Eindhoven, has explored the potential for more sympathetic imperfection in automated processes, from

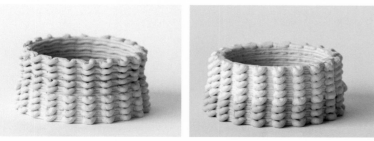

Variations in materials, pressure, speed, and printing diameter generate craft-like variation

David Bowen
Growth Modeling Device

→ #materials → #sustainable

Challenging the Attempt to Conquer Nature through Reproduction

Year: 2009
Client/Purpose: Artwork
Partners: Bemis Center for Contemporary Art, The University of Minnesota, Duluth
Technique: Fused deposition modeling
Printer/Service: RepRap
Material: ABS plastic
Production time: 1 year
Budget: $8,000
Edition: 1

In the Western world, nature has historically been represented through two different motifs: the classical involved soft, visually pleasant forms reminiscent of the bountiful Garden of Eden, while the picturesque made use of dramatic staging and jagged forms to inspire a mixed sense of fascination and apprehension. As scientific knowledge of the natural world increased, however, the metaphorical and symbolic elements of nature have largely been transcended by documentation of the self-regulating complexity of natural systems. Still, visual depictions of this complexity have maintained an intriguing sense of wonder, from the detailed drawings of botanists to time-lapse videos of plants sprouting and flowers blooming. Growth Modeling Device translates our desire to understand and archive natural processes into an automated process. David Bowen's machine brings together a 3D scanner, a 3D printer, a conveyor belt, and an onion plant. Every 24 hours, the plant is scanned from one of three angles and replicated in real time as a flattened silhouette; once complete, the model is advanced 17 inches along the conveyor to make room for the next day's reproduction. The machine juxtaposes the organic intricacy of the real, living plant with the rather sterile two-dimensional reproductions, questioning the validity of any replica of the natural world, even if our tools have become more precise and objective.

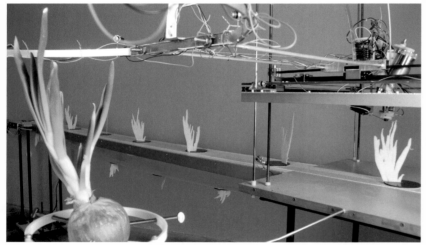

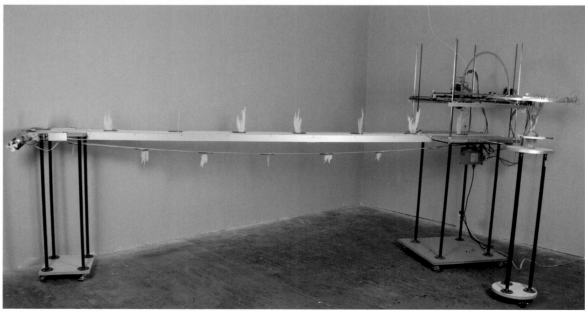

The onion plant is scanned daily and the results are printed as a two-dimensional profile

mischer'traxler

'till you stop – cake decoration

→ #machines

Exploring the Limits of Desire and Ornament

Year: 2010
Client/Purpose: For exhibition
Partners: MAK Vienna
Technique: Paste extrusion (sugar)
Printer/Service: N/A
Material: Sugar glazing
Production time: 2 months
Budget: $1,000
Edition: 1

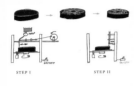

STEP I   STEP II

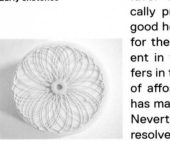

Early sketches

A finished cake

In his formative 1913 essay "Ornament and Crime," Adolf Loos criticized the excesses of decoration and visual styling of turn-of-the-century applied arts as an immorally self-indulgent waste of time and material. In declaring that "the evolution of culture is synonymous with the removal of ornament from utilitarian objects," he anticipated the later marginalization of laborious manual work in favor of increased automation, which theoretically predicted a greater diffusion of education, good health, and employment of a higher intellect for the masses. The current reality is very different in those respects, of course, but it also differs in terms of ornament. In fact, the proliferation of affordable technology for mass customization has made ornamentation more efficient than ever. Nevertheless, the question of taste remains unresolved: just because we can add as much detail and decoration as we want using new methods of rapid prototyping, is it advisable? Vienna-based design studio mischer'traxler asks this very question with "till you stop," a cake decoration machine that applies a spirograph pattern of pink sugar glazing and drops silver sugar pearls onto the surface of a cake. The amount of decoration is subject to the user's discretion: when the cake looks satisfactory, a button is pressed to stop the process. The project speculates on the aesthetic and ethical dilemmas raised by the increasing mobilization of productive apparatus to satisfy our every whim.

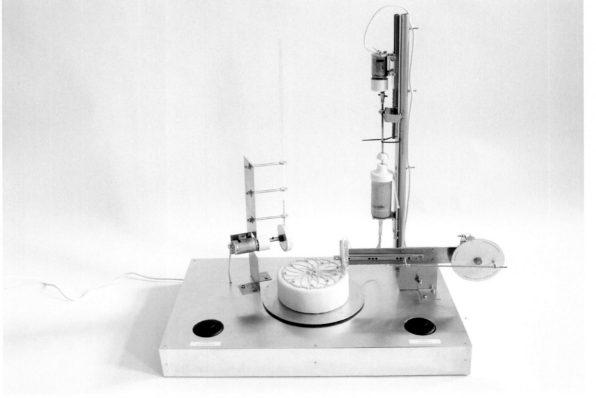

The machine decorates the cake with glazing and silver pearls until the user presses the button

DUS Architects
The Potato Eaters

→ #sustainable → #materials

Potatoes on – and in – Parametric Tableware

Year: 2013
Client/Purpose: Research
Partners: Future Food House, Stichting aan de Middendijk
Technique: Fused deposition modeling
Printer/Service: Ultimaker
Material: Potato starch, additives
Production time: N/A
Budget: N/A
Edition: In development

Though one class of 3D printers may be described as low-cost, they are still expensive enough to discourage a certain amount of material experimentation; the low tolerance of the precise modeling technique may be a contributing factor to the

The Potato Eaters series thus embeds the very substance of tableware items with the same food they will eventually contain.

almost total prevalence of standardized, mass-produced materials, from ABS plastic to white nylon. Against this background, The Potato Eaters project by DUS Architects is notable not only for its utilization of a new organic substance but also for its attempt to find a material that can be manufactured locally from leftover materials. Referencing the humble loop of work and sustenance displayed in the eponymous Van Gogh painting, the project processes leftover potato parts, which could be harvested from the popular French fry stands found around the Netherlands, to create a potato starch composite. This material is then fed into a 3D printer to produce a tableware collection. Each of the pieces in the series is unique, modeled algorithmically around five fixed points in space to generate a parametric form like a potato.

Matthew Plummer-Fernandez
Disarming Corruptor

Year: 2013
Client/Purpose: Research
Partners: N/A
Technique: N/A
Printer/Service: N/A
Material: N/A
Production time: N/A
Budget: None
Edition: Free download

→ #authorship

Encoding Objects for Undetectable Transfer

On May 5, 2013, Defense Distributed published the virtual model for the Liberator, the first fully 3D-printable gun, via DEFCAD, an online database initially created to host files banned from Thingiverse. Four days later, the U.S. Department of State requested that the files be taken down, although they can still be found on file-sharing web sites. While there is currently no action being taken against those who distribute or download the files, the tendency of the online community to create new modes for the transmission of information (including copyrighted and illegal material) is generally accompanied by increased tracking, surveillance, and (at times) active response by governments. Using the Processing language, Matthew Plummer-Fernandez has designed the software Disarming Corruptor as a way of making 3D files completely unrecognizable while they are being transferred in order to evade notice. Inspired by rotor cipher machines like the Enigma (first developed in the 1920s), the application corrupts virtual models by torqueing the 3D mesh and adding glitches. The recipient of the file can then use the same software and a unique seven-digit code to reverse the damage, revealing the original file (an incorrect code would only impose further damage on the file). Disarming Corruptor would thus enable the dissemination of digital contraband in "plain sight."

Driessens & Verstappen
Accretor

Year: 2012 to 2013
Client/Purpose: Artwork
Partners: N/A
Technique: Polyjet
Printer/Service: Objet500 Connex
Material: VeroBlack acrylic photopolymer
Production time: N/A
Budget: N/A
Edition: N/A

→ #complexity → #software

Aggregating Form in Generative Processes

Amsterdam-based artists Erwin Driessens and Maria Verstappen first became interested in generative processes in response to the self-powering machine of the late twentieth-century cultural institution and its implications for the production of meaningful content. By writing scripts for the creation of artwork rather than making the art themselves as fallible humans, they wanted to explore the chaos and spontaneous order underlying the cultural sphere. Alongside experiments in casting icicles, carrots, and potatoes, eroding pewter objects, and melting beeswax, they also ventured into generative systems in the virtual sphere, realizing several projects using 3D printing. While the Breed series (1995 to 2007) represents a level of complexity that is manageable enough to be assembled by hand, Accretor pushes the detail and algorithmic variability to an incredible degree. Driessens and Verstappen wrote this software to invent forms through accretion, composing a set of rules to guide the build-up of granule on top of granule. Finally, a few sculptural iterations were printed in acrylic resin at extremely high resolution. The process takes place with practically no intervention of the human hand, yet the results are curiously organic in their intricacy.

Accretor #1758-2

Accretor #2777-4

Accretor #5188-3

Yahoo! Japan
Hands On Search (Sawareru)

→ #empowerment   → #interact   → #machines

Making the Visual Tangible for the Visually Impaired

Year: 2013
Client/Purpose: Education
Partners: Hakuhodo Kettle, AID-DCC, katamari, University of Tsukuba
Technique: Fused deposition modeling
Printer/Service: MakerBot Replicator 2
Material: PLA bioplastic
Production time: 5 months
Budget: N/A
Edition: 1

In a small school in Tsukuba, Japan, a child stands in front of a large white machine with two plates of glass recessed in marshmallow-like protuberances. One of them is a computer screen, while the other is the window to a 3D printer. The child calls out a word, and within a matter of minutes its shape is built up inside the machine. Far from being the latest novelty toy, this machine represents a very serious endeavor: the child is blind, and the school is the Special Needs Education School for the Visually Impaired. The machine, called Hands

On Search, is the outcome of research initiated by Yahoo! Japan into the potential of 3D printing to support haptic or kinetic learning processes for those who cannot see. Hands On Search demonstrates how the on-demand principle of 3D printing can be harnessed by larger social systems such as education, which once had a one-size-fits-all mentality, in order to better serve their aims. Not only does the machine equip the visually impaired with the open-ended power of search, but it also generates an immediate physical manifestation of that search term based on a visual analysis of the results, with data contributed by Nissan, amana, and Tokyo Sky Tree, among other private entities. If no 3D model is found, the network is prompted to build up a library of available forms using an advertising program from Yahoo! Japan called Links For Good.

Children activate the machine by speaking words that are then 3D-printed

Bengler

Terrafab

→ #customization → #open source

Mutually Inclusive Ownership for Norway's Open-Source Landscape

Year: 2013
Client/Purpose: For sale
Partners: Norwegian Mapping Authority
Technique: Inkjet powder printing
Printer/Service: Shapeways
Material: Colored gypsum
Production time: 1 month
Budget: $10 to $2,500 per print
Edition: On-demand

The open source movement began as an initiative rooted in the private enterprise of inventors, from Benjamin Franklin's myriad creations to Linus Torvald's Linux operating system. Despite some resistance, the productivity of the commons has slowly begun to pervade official institutions that traditionally safeguarded their data. In 2013, before the centrist-left coalition ceded control of Norway's cabinet to the conservative party, they mandated a partial release of the geospatial data held by the Norwegian Mapping Authority into the public domain. To mark the occasion, the government agency asked Bengler, an Oslo-based studio renowned for their information design projects, to assist in the release. Rather than pursue a digital visualization, however, they explored a way of physically manifesting the data as something that any individual could sample. Terrafab is an online platform that allows users to isolate any square-shaped section of the Norwegian landscape and generate a 3D elevation model, including texture mapping to represent water, vegetation, and snow. The model can then be downloaded or sent to Shapeways to be printed in full color. Bengler's project demonstrates a fundamental principle of the open source approach: shared data can generate products for individual ownership, yet the nature of that propriety remains completely inclusive and overlapping.

Theo Jansen
3D-Printed Strandbeests

→ #authorship → #assembly free

Year: 2011 to present
Client/Purpose: For sale
Partners: Bo Jansen, Tim van Bentum
Technique: Selective laser sintering, polyjet
Printer/Service: Shapeways
Material: Polyamide
Production time: 10 months
Budget: $70 to $140 per print
Edition: On-demand

A Technological Genesis of Miniature Wind-Powered Animals

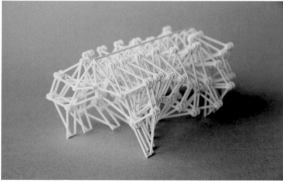

Dutch artist Theo Jansen is internationally renowned for his Strandbeest, a "new form of life" that emerged in 1990, comprising structures made of wood, plastic tubing, and sails that can harness the wind, enabling the creation to walk along beaches with no additional power source. While Jansen's models measure several meters in height, they have also inspired an exploration of the potential presented by rapid prototyping at a miniature level. Product designers Bo Jansen and Tim van Bentum worked to recreate the intricate mechanisms for motion using the printing technologies offered by the Shapeways printing service: the first model, Animaris Geneticus Parvus, was released in April 2011 with 74 interlocking, moving parts that were printed in a single pass. Since then, the 3D-printed Strandbeest "genus" has continued to grow, with Ondularis, Gracilis, and Larva models in addition to the Parvus. These constructions, most of which are rendered in laser-sintered nylon or multijet modeling for the smallest versions, attempt to replicate the drama of the originals at a small scale. More important, they also allow for rapid adaptation and "genetic" modification, creating a digitally enhanced methodology for the evolution of new forms of the iconic Strandbeest.

F.A.T. Lab – Free Art and Technology
The Free Universal Construction Kit

→ #empowerment   → #authorship   → #building blocks   → #open source

Year: 2012
Client/Purpose: Research
Partners: Sy-lab.net
Technique: Polyjet
Printer/Service: Objet500 Connex
Material: Acrylic photopolymer
Production time: 3 months
Budget: $1,000
Edition: On-demand

Open-Source Bridges between Competing Copyrights

The complete Free Universal Construction Kit

The Free Art and Technology Lab, a collective based in New York City, tends to make reference to light-hearted themes from popular culture that disguise incredibly serious societal and political questions, from a plug-in that removes all traces of Justin Bieber from the Internet to a cradle-rocking robotic arm for busy parents. In the same vein, the Free Universal Construction Kit may have an ostensibly playful and practical function, but it is nevertheless a considered meditation on the illogical barriers to free expression created by mainstream industry and copyright laws. The kit works as an interface between ten popular commercial construction toys such as Lego, Tinkertoy, Lincoln Logs, and so on, and is designed as a matrix of nearly 80 two-way adaptors that allow for hybrid constructions. F.A.T. Lab used an optical comparator to measure and model the various pieces at a precision of one ten-thousandth of an inch (much finer, in fact, than a desktop 3D printer could achieve) to create a perfect snap fit. As a result, the Free Universal Construction Kit reveals a dimension of interoperability that resides in most open-source platforms but is completely alien to the world of commercial manufacture: indeed, this kit would never be made by any one company, since it would inherently lead consumers to buy products from other providers. Additionally, the kit must remain a provocation – the virtual model may be distributed online, but it cannot be sold due to its appropriation of proprietary protocols.

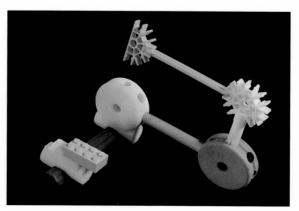

Example of a hybrid construction

DesignLabWorkshop and Brian Peters
Building Bytes

Year: 2012
Client/Purpose: Research
Partners: Europees Keramisch Werkcentrum
Technique: Paste extrusion (ceramic)
Printer/Service: Modified
Material: Earthenware
Production time: 2 months
Budget: N/A
Edition: N/A

→ #building blocks → #architecture

A New Architectural Language of Custom Ceramic Bricks

The material and scalar abstraction of the 3D printing process is often seen as a drawback, removing the grain and creative "friction" of more analog design processes. This viewpoint, however, fails to consider the advantages created by such an abstraction, including its ability to act as Ionia, and developed the project during a residency at the Europees Keramisch Werkcentrum, one of the world's most specialized facilities for the realization of technically challenging art and design objects. Building Bytes treats 3D printers as "portable, inexpensive brick factories," imagining

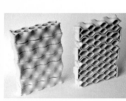

Honeycomb brick wall prototype

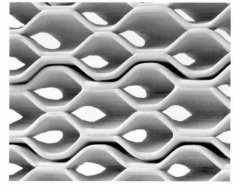

Honeycomb brick detail

a shared language between very disparate spheres of production. One example is Building Bytes, a series of modular ceramic "bricks" for building-scale constructions designed by Brian Peters, the founder of Ohio-based DesignLabWorkshop. Peters drew on his education at the technologically progressive and highly theoretical Institute for Advanced Architecture of Cata- different scenarios – including domes, columns, stacked walls, and vertical tiles – made possible by four brick types, including honeycomb, ribbed, interlocking, and X-shaped variations. Building Bytes are also engineered to be lightweight and to distribute stresses using locally available liquid material, such as concrete or earthenware, hypothesizing a new model for self-produced architecture using complex components with simple, traditional construction techniques.

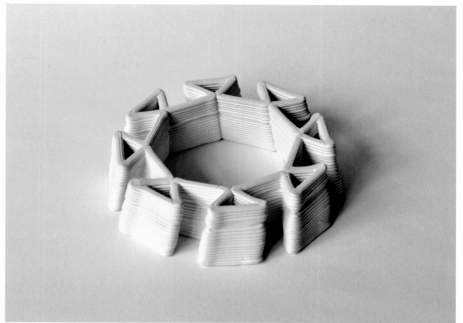

Ribbed bricks are ideal for constructing columns

ROB Technologies
ROB-Made Fabrication Method

→ #building blocks → #machines → #architecture

Intricate Constructions with a Robotic Arm

Year: 2006 to present
Client/Purpose: Research
Partners: ETH Zurich,
Gramazio & Kohler,
Keller Zeigeleien
Technique: Brick deposition
Printer/Service: R-O-B
Material: Bricks, glue
Production time: Varied
Budget: N/A
Edition: N/A

The prevailing vision of the 3D printer is that it is a solitary object sitting on a fab lab bench or on the desk of a forward-thinking hobbyist. These scenarios, however, seem somewhat limiting in comparison to a larger network of automated and semi-automated processes and equipment, all working in tandem. One of the best examples is R-O-B, a mobile fabrication unit that can build complex curved walls, columns, and other architectural structures according to computer instructions. Developed by ETH Zurich, Gramazio & Kohler, ROB Technologies and Keller Zeigeleien, the robot has been creating intricate structures since 2006, from the undulating wall panels of the Gantenbein Winery to the delicate lattice of the Ofenhalle Pfungen. 3D printing, in the traditional sense, could be seen in the automated application of glue to the bricks – yet the entire construction could equally be seen as a large system of additive manufacturing, depositing identical units (bricks) rather than a liquid substance. R-O-B is a modern interpretation of the traditionally labor-intensive craft of bricklaying in a contemporary context, including a symbiotic collaboration between people and machines and the optimization of structures – specifically the increased stability of the walls thanks to their curvature. Furthermore, the parametric automation makes possible new qualities of light and air transmission that would have been prohibitively expensive in the past.

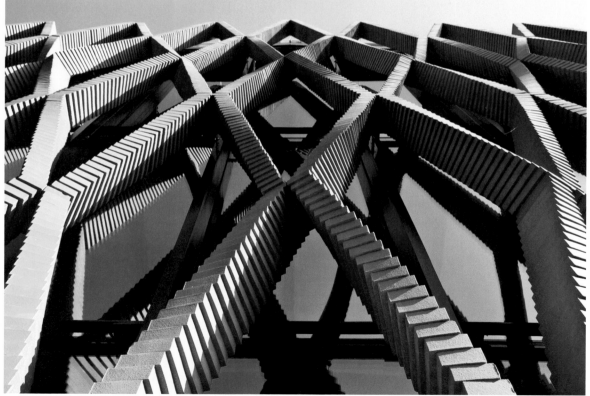

Complex structures are achieved by automating the application of glue and stacking of bricks

Joris Laarman Lab, Petr Novikov and Saša Jokić
Mataerial

→ #machines → #materials

Unleashing the Printer from its Structural Confines

Year: 2013
Client/Purpose: Student research
Partners: Institute for Advanced Architecture of Catalonia (IAAC)
Technique: Anti-gravity object modeling
Printer/Service: ABB 2400L robotic arm with S4 controller and custom nozzle and extruder
Material: Two-component thermoset polymer
Production time: 5 months
Budget: $11,000
Edition: In development

Despite the rapid explosion of new materials, processes, and resolutions in the general category of 3D printing, the printing bed itself continues to exert a limiting factor on the possibilities that can emerge. As a rule, the printing apparatus must be able to move around on a computerized rig covering the extents of the printable space; as may be expected, increases in the size of this set-up are difficult to achieve while maintaining the degree of precision needed for successful results. Rather than attempt to do this, however, designers Petr Novikov and Saša Jokić rethought the process entirely. During their time at Joris Laarman Lab they envisioned the printer head as an independent object moving freely in space, without the need for a support structure. The result was Mataerial, a robotic arm that emits a special thermoset polymer which hardens in midair. Mataerial can work using the conventional additive manufacturing techniques of slices and coils, but it is also able to draw complex curved lines through space, in a technique the pair have attempted to patent as anti-gravity object modeling. Mataerial radically alters the concept of 3D printing, freeing it from its status as a process that must take place in a tightly controlled, precisely defined environment and establishing it as a method for spatial intervention and large-scale manufacture that can occupy and interact with real-life environments.

Color printing

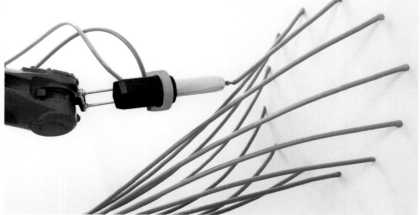

Mataerial can print on any surface, at any angle

Studio Smith|Allen

Echoviren

Year: 2013
Client/Purpose: For exhibition
Partners: Project 387, Type A Machines
Technique: Fused deposition modeling
Printer/Service: Type A Machines Series 1 3D Printer
Material: PLA bioplastic
Production time: 3 months
Budget: $1,200
Edition: 1

→ #building blocks   → #architecture

An Ephemeral Pavilion Assembled in Nature

Contemporary architecture is ingrained with a certain paradox: those who study it often learn much more advanced software and experiment with more complicated forms than they will ever use in real-life projects, but the scale of a building makes it extremely difficult for them to approach these ideas on their own, in the way a designer works with prototypes. Not long after architects began to 3D-print quick, detailed scale models, however, the idea to 3D-print an entire building arose. The challenge has been approached in different ways, generating a variety of declared "firsts". Echoviren, the pavilion created by Oakland-based practice Smith|Allen, is one example: here, the designers chose to produce nearly 600 unique "bricks" on a fleet of seven off-the-shelf personal 3D printers, not only to prove the viability of 3D-printed architecture but also to demonstrate a more functional output from these cheaper machines. The parts were connected with a paneled snap fit, building up a curving, web-like wall that was assembled in a Californian redwood forest in four days. The pavilion was also designed to decompose naturally: printed in PLA bioplastic, the structure will eventually become a microhabitat for living organisms.

EFFALO
domekit 3D-Printed Connector

Year: 2011
Client/Purpose: Research
Partners: Michael Felix,
Emily Felix, Robby Kraft,
Andrew Kurtz, Ezra Spier,
Mark Cohen
Technique: Fused deposition
modeling
Printer/Service: Modified
MakerBot Cupcake
Material: ABS plastic
Production time: 3 years
Budget: $7,000
Edition: In development

→ # building blocks   → #empowerment   → #architecture

A Geodesic Dome in the Age of Networked Fabrication

Early prototypes

**domekit 3D-printed
connectors**

Dome with domekit 3D-printed connectors

The theoretical principle of 3D printing has always stretched beyond the practical constraints on the size and efficiency of the printers available at the time. As contemporary architecture has cultivated an interest in parametric design and mass customization, it has also begun to experiment with rapid prototyping. One approach to the limitations of size, exemplified by domekit, a project by the Asheville-based collaborative EFFALO, has been to 3D-print only the joints and to connect them using standardized tubes or rods cut to length. While many geodesic domes have been designed since they were first popularized in the mid-twentieth century, the EFFALO project has evolved in a particu-

larly twenty-first-century paradigm, including the crowdsourced fabrication of original prototypes using the MakerBot mailing list, modified ball-joint files found on Thingiverse, and financing through Kickstarter. Although additive manufacturing may be a time-consuming way to produce identical joints for one dome, it offers a variety of benefits for the production of many different kinds of joints for domes, be they small, large, more angular, or more spherical. This customization has been further facilitated by the release of a domekit iOS app that can generate parametric part lists and assembly diagrams. The designers have also invested their experience in crowdsourced 3D printing into the development of MAKERFACTORY, an online service to connect demand with CNC machines around the world.

Parts required to make the dome

Dome with domekit 3D-printed connectors

Dome made for a music video using 3D-printed connectors

Rael San Fratello Architects
Saltygloo

Year: 2013
Client/Purpose: For exhibition
Partners: The University of California Berkeley, San Jose State University, Museum of Craft and Design
Technique: Salt printing
Printer/Service: 3D Systems ZPrinter 310 Plus
Material: Salt
Production time: N/A
Budget: N/A
Edition: 1

→ #materials → #sustainable → #architecture

Printing Architecture with Naturally Harvested Salt

3D printing is referred to as a "zero waste" process, but it nevertheless relies on a strict palette of materials limited almost entirely to plastic. Furthermore, while the corn-based bioplastic polylactic acid is becoming popular for fused deposition modeling, it is debatable whether its total environmental impact is less than that of petroleum-based plastic (moreover, it is a highly processed material). While plastic may be suitable for small objects in both additive manufacturing and traditional methods of manufacture, it is not a viable solution for large-scale architectural works. The Oakland-based architecture studio Rael San Fratello founded Emerging Objects in 2012 in response to this very issue, initiating experiments in rapid prototyping with innovative materials, and often using a natural material or waste product as a substrate combined with an appropriate binder. Saltygloo, for instance, consists primarily of salt from San Francisco Bay, harvested in crystallization ponds using natural sun and wind power. The 3D printer binds the salt with glue, resulting in a material that is strong, waterproof, and inexpensive. Rael San Fratello first presented this material in the form of a pavilion created for an exhibit at San Francisco's Museum of Craft and Design. Inspired by the Inuit igloo, the semi-structural shell was made from 336 unique printed panels and tensile aluminum rods, demonstrating the translucency, low weight, and formal potentials of the material.

Digital Grotesque
Printed Space

→ #architecture → #complexity

Resurrecting the Rococo with Rapid Prototyping

Year: 2013
Client/Purpose: Research
Partners: Michael Hansmeyer, Benjamin Dillenburger, ETH Zurich, voxeljet, Fonds régional d'art contemporain
Technique: Sand printing
Printer/Service: voxeljet VX4000
Material: Quartz sand, resin, pigments, alcohol, shellac
Production time: 1 year
Budget: N/A
Edition: N/A

1:10 scale prototype - gold

Initial coating test

Gilding

3D printing is often invoked in discussions about rethinking architecture for contemporary terms of production, but the large scale of buildings makes much of its potential difficult to explore. For example, though additive manufacturing can produce unique objects with every print at no additional cost, most large-scale experiments to date have involved a form of standardized interlocking panel with a limited degree of variation. Unsurprisingly, the labor in the design process remains a substantial cost for additive manufacturing, encouraging recourse to standard forms; meanwhile, the architecture that most successfully embodied complexity in history – the religious and monu-

mental works of the Baroque – were created when cheap labor could transform very expensive material into stunningly detailed constructions. Digital Grotesque is the first experiment to reverse this condition in the pursuit of a detailed, ornamental aesthetic that is unique at each point in the design. The research group based at ETH Zurich used a computational design strategy to transform a simple wall surface into 260 million individual facets. This geometry was then realized in 3D-printed sandstone, creating an incredibly sculptural room from 64 blocks. While the surface was heavily refined after printing to enhance its finish and durability, the assembly took only one day. Digital Grotesque has thus reintroduced an aesthetic that became largely impossible during modernism as a viable design approach.

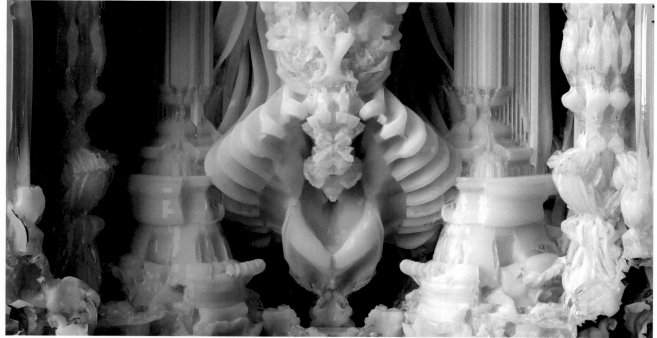

Design development of the grotto

Grotto assembly

Grotto interior, detail

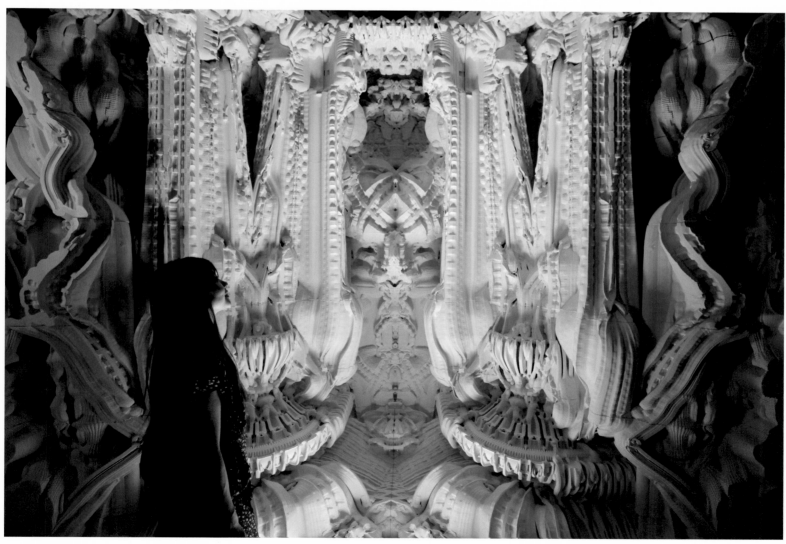

Grotto interior

A printed model of the Bergkerk, seen from the altar

Driessens & Verstappen
Solid Spaces

→ #3D scanning   → #software

Understanding How Machines See the World

Year: 2013
Client/Purpose: Artwork
Partners: N/A
Technique: Polyjet
Printer/Service: Objet500 Connex
Material: VeroGray acrylic photopolymer
Production time: N/A
Budget: $2,300
Edition: 1

In line with their interest in automated or generative processes of creating art, Driessens & Verstappen created the Solid Spaces series as an exploration of the disparity between digital syntax

A printed model of the Bergkerk, seen from the aisle

A pan-tilt laser scanner measures the Bergkerk in Deventer

and human thought processes. The works are created by plotting out an interior gallery space, from the Bergkerk in Deventer to Berlin's DAM Gallery, using a pan-tilt scanner with a laser-sensor. The digital model it builds up is then printed to scale and displayed in situ. Solid Spaces highlights the productive potential of mechanical fallibility: when the human eye scans such spaces, we automatically interpret and fill in missing information to assemble a reasonably convincing mental concept of the space, whereas a computer has no such inclination unless it is directed to interpolate the data. Instead, the scanner's incomplete data renders the space as a point cloud, resulting in strange artifacts such as an infinitely long window volume and fan-shaped gaps emerging behind visual obstructions. By showing these models in the same space that informed their emergence, Driessens & Verstappen imply a growing mutual intelligibility between living beings and electronic devices. In a future phase of the project, the digital scanner will also acquire the ability to learn through a feedback loop of observation, memory, and interpretation, and thus communicate through a different language of space.

A three-dimensional point cloud of spatial readings

DUS Architects
KamerMaker

Year: 2012 to present
Client/Purpose: Research
Partners: Ultimaker, Rooie Joris, Xtrution, Doen Foundation, Dutch Creative Fund, Fiction Factory
Technique: Fused deposition modeling
Printer/Service: KamerMaker based on the Ultimaker
Material: Various bioplastics
Production time: 2 years
Budget: N/A
Edition: In development

→ #architecture → #building blocks → #machines

3D-Printed Architecture for Everyday People

3D printing has fascinated architects since its birth as a way of producing scale models, but the combination of unlimited formal possibilities and the limited potential for full-scale realization in the near future has tended to fuel the fantasies of many designers toward a self-consciously extravagant visual language. The Dutch studio DUS Architects has a different agenda: they are focused on bringing the benefits of additive manufacturing on a large scale to a mass audience, and further involving them through processes of co-creation. Accordingly, their KamerMaker ("RoomBuilder" in Dutch) is a scaled-up Ultimaker 3D printer, six meters in height, standing in a public park in the center of Amsterdam, where it is incrementally building the parts for a canal house in various bioplastics. At the moment, the KamerMaker's build

envelope is $2 \times 2 \times 3.5$ meters, and a second machine is currently in development. DUS Architects envision this massive 3D printing both as a way of democratizing conventional architecture and as a prototype for on-demand shelters in emergency situations. Furthermore, it is a zero-waste process that handles raw material directly on-site, avoiding the need for the secondary shipment of components for intermediate processing. The KamerMaker introduces the concept of physical experimentation in the building process – an option that has rarely been available in the past due to the high cost of construction.

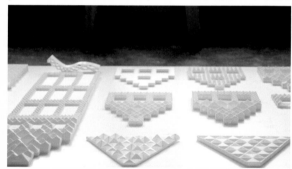

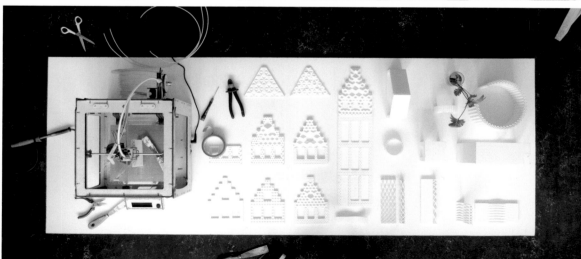

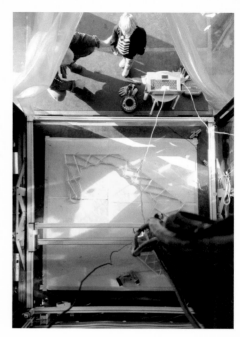

Rendering of the planned 3D-printed Canal House

Markus Kayser
Solar Sinter

Year: 2008 to present
Client/Purpose: Student research
Partners: Royal College of Art
Technique: Selective laser sintering
Printer/Service: SolarSinter
Material: Sand
Production time: 3 years
Budget: N/A
Edition: 1

→ #new industries　→ #machines　→ #materials　→ #sustainable

Making Objects from Sand and Sunlight

The idea of distributed manufacturing as an outcome of 3D printing is theoretically powerful in terms of altering the normative hierarchy of consumption and production in a globalized but unequal world, and the spread of fab labs across 35 countries, from Afghanistan to Suriname, promises an increased diffusion of knowledge in relation to CNC technologies. Still, the system is best adapted to contexts with abundant electricity and easy access to standardized materials and electronics. With SolarSinter, however, German designer Markus Kayser has invented a system for places that lack such infrastructure but have an abundance of sand as a raw material and the energy of sunlight. The SolarSinter exploits the collimated light channeled by a sun-tracking Fresnel lens (in which scattered light forms an infinitely focused beam of parallel rays) to trace lines in a bed of sand, raising the temperature above 1,400°C in order to melt the silica sand particles, creating glass. The machine is controlled by a computer with a battery charged by photovoltaic panels for continuous production. After several hours of printing and cooling, the object can be dug out of the sandbox, revealing a rough exterior and a hard glass upper surface. Kayser's creation, first tested in the Sahara Desert in 2011, predicts an increasing liberation of the apparatus of manufacture from centralized systems of control, taking advantage of open-source innovation.

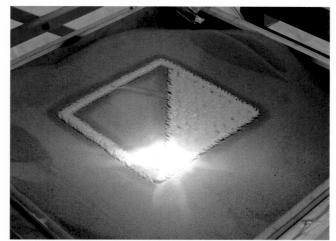

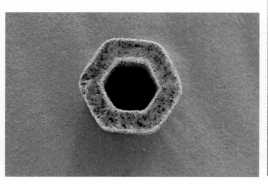

Tethers Unlimited
SpiderFab

→ #building blocks

Self-Manufacturing Infrastructures in Space

Year: 2011 to present
Client/Purpose: Research
Partners: NASA Advanced Innovative Concepts
Technique: SpiderFab
Printer/Service: N/A
Material: High-performance thermoplastic and carbon fiber composites
Production time: N/A
Budget: $500,000 NASA innovation grant
Edition: In development

The inventions associated with space exploration have historically been subject to a figurative law of gravity – they tend to come back to Earth and permeate everyday terrestrial life in the form of objects such as cordless power tools, memory foam, scratch-resistant glass, and freeze-dried food. In the case of 3D printing, however, a technology designed for small-scale prototyping is now being radically repositioned as a critical innovation for the realization of large-scale infrastructures in space. Rather than assembling satellite components on Earth, folding or compressing them, and sending them into orbit, Seattle-based Tethers Unlimited is suggesting that we deploy self-assembling satellites equipped with 3D printers to build their own components in situ. The company is in the process of patenting a technique called SpiderFab, which combines optimized carbon fiber placement with fused filament fabrication using high-performance thermoplastics. Tethers Unlimited envisions an enormous increase in the possible magnitude of solar infrastructures, from several dozen meters in length to a kilometer or more, with an associated increase in solar power collection and satellite bandwidth, resolution, and sensitivity for a variety of functions.

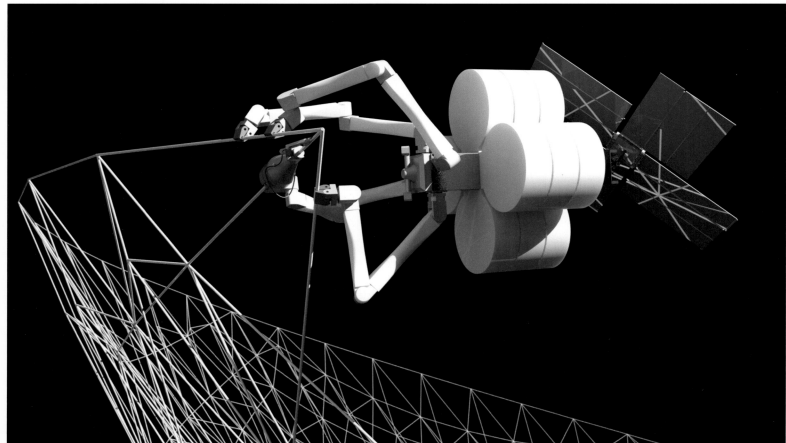

3D printing solar arrays in space would allow for much larger, more complex structures

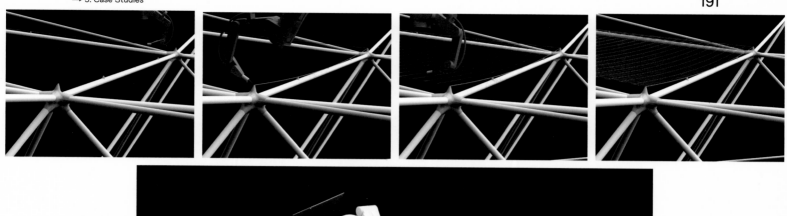

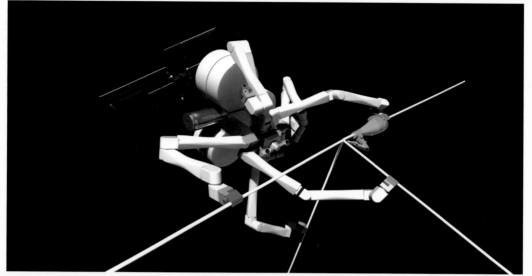

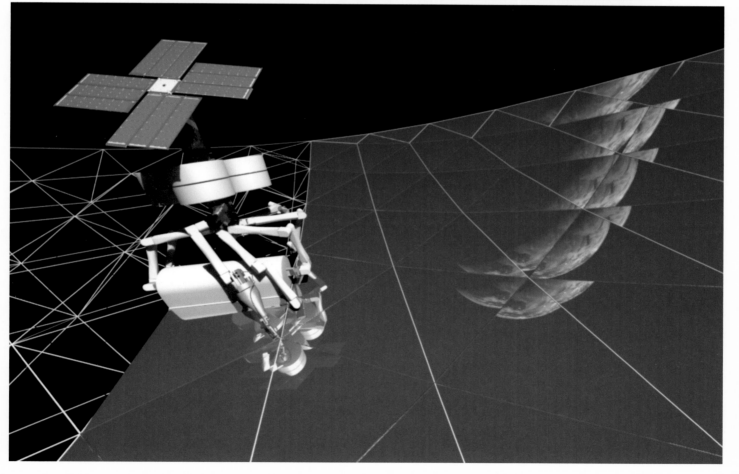

Foster + Partners
Habitable Lunar Settlement

→ #architecture   → #machines   → #materials

Printing Moon Colonies with Lunar Soil

Year: 2013 to present
Client/Purpose: Research
Partners: European Space Agency, Alta, Monolite UK, Laboratory of Perceptual Robotics (Scuola Superiore Sant'Anna)
Technique: Selective laser sintering, inkjet powder printing
Printer/Service: EOS P390, D-Shape
Material: Regolith (lunar soil)
Production time: Ongoing
Budget: N/A
Edition: N/A

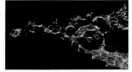

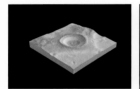

As one of the most automated and flexible forms of manufacture, 3D printing has drawn particular interest from the field of space exploration and possible colonization. Since the cost of transporting any substantial quantity of building materials in a spaceship would be enormously high, another solution for construction has to be found if humans are to reside on the moon in large numbers. The European Space Agency has set up a consortium that includes architects Foster + Partners and Monolite UK, the 3D printing company behind the D-Shape large-scale stereolithography machine, to explore this very problem in terms of additive manufacturing. The group is studying the use of regolith, or lunar soil, as the substrate for the printing process, in combination with chemical binders. Expanding on a concept pioneered by architect Wallace Neff in 1942 for mass housing in developing contexts, the architectural strategy uses inflatable domes as temporary support structures for the construction of a base for four people. In order to reduce the printing time and the

Autonomous robots are used to 3D-print a cellular structure that protects the inhabitants from gamma radiation, meteorite impacts, and extreme temperature fluctuations

A lunar outpost near the moon's south pole

amount of binder necessary, the architects have designed the walls as a foam-like mesh of hollow, closed cells that is still strong enough to protect the inhabitants from meteorites, radiation, and temperature fluctuations. The group has already printed a mock-up weighing 1.5 tons, simulating lunar soil as well as environmental conditions, and continues to optimize the speed with the aim of printing an entire building in one week.

Creating a foam-like structure with simulated lunar soil on the D-Shape printer

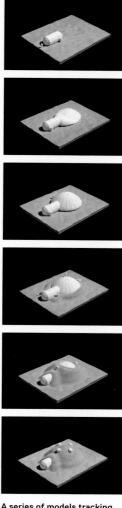

A series of models tracking the 3D printing process

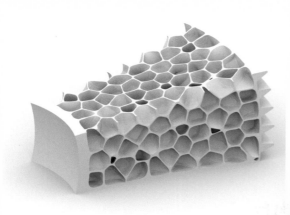

The foam-like structure is designed for maximum strength and minimum mass

A 1.5-ton building block printed as a case study

SHAPES iN PLAY
infObjects

Year: 2011
Client/Purpose: Research
Partners: Hochschule Coburg,
EOS Electro Optical Systems
Technique: Selective laser
sintering
Printer/Service: N/A
Material: Polyamide
Production time: 6 months
Budget: N/A
Edition: 1

→ #customization   → #software   → #intangible

Designing Tableware with Parametric Analysis

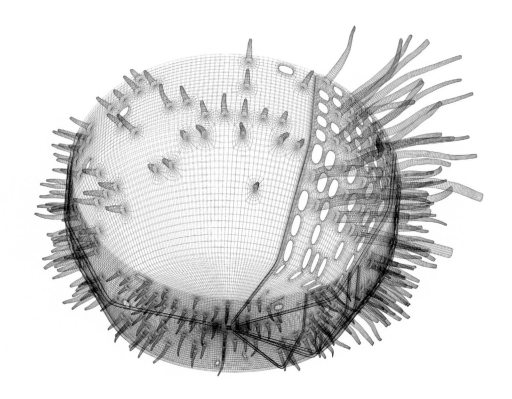

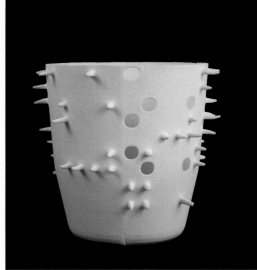

One of the most significant problems of the modern period is the massive increase in available data and the resulting demand for tools to filter and interpret this information. From a visual perspective, software like Processing has been essential for the proliferation of customized maps and infographics on mainstream media channels. Using the same software, the Berlin studio SHAPES iN PLAY has translated the desire for contextual information into a physical result with infObjects, a tableware series that attempts to analyze and communicate the environmental, nutritional, and financial footprint of the food items it contains. The designers began with a set of archetypal shapes – a cup, a bowl, and a plate – and then adjusted these forms based on the parameters associated with the individual components of different foodstuffs. For example, carbon dioxide emissions were correlated to holes in the surface of the vessel, while calorie content was shown as roots, and the price of the ingredients determined the height of the respective segments. When produced for a variety of dishes, the tableware allowed people to make an informed comparison between the consequences of each food choice. infObjects are not intended to serve as functional objects, but they are also relative: a food item with a large carbon footprint or a very high calorie count would make the vessel more precarious and perforated and thus even less functional.

Caipirinha

Eine Portion ~ 170 g
1264 kJ
223 g CO2e
140 Cent

Schoko-Sahne Pudding

Eine Portion ~ 195 g
2049 kJ
260 g CO2e
52 Cent

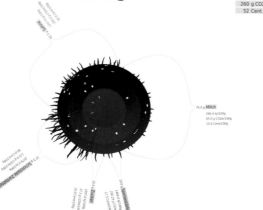

100 g Eier

646 kJ

22 Cent

480 g CO₂

The nutritional, environmental, and economic impacts of different foods are analyzed

Studio Maaike Roozenburg
Smart Replicas

→ #interact   → #intangible

From Irreplaceable Artifact to Virtual Reproduction

Year: 2010 to 2013
Client/Purpose: Research
Partners: Delft University of Technology, Delft Heritage, Museum Boijmans Van Beuningen
Technique: Stereolitography (Digital light processing photopolymerization)
Printer/Service: Mareco Prototyping
Material: Gypsum, nylon, vinyl, porcelain
Production time: 3 years
Budget: N/A
Edition: N/A

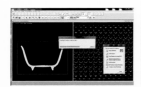

Smart technologies allow more interactive learning with the objects

Digital model of porcelain cup

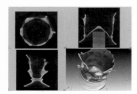

Glasses analyzed with a CT scanner

Digital reconstruction of a seventeenth-century glass

Digital reconstruction of a porcelain cup

3D prints in plaster, vinyl, and plastic/work in progress

Shaped by Enlightenment-era values, museums have historically engaged in the systematic acquisition of manmade objects in order to foster a greater understanding of human society. This aim prioritizes the conservation of rare items, but by necessity also limits the extent of visitors' interaction with them. For the past several years, however, designer Maaike Roozenburg has been developing an experimental approach toward the collection of the Museum Boijmans Van Beuningen using new technologies. Her Smart Replicas are made by using CT scans to generate incredibly precise 3D models of historical artifacts, such as teacups. Then, using extremely high-resolution rapid prototyping and augmented reality technology, she is able to recreate the originals as objects that can be handled by visitors and used to visualize a large body of archaeological and ethnographic research. Roozenburg's project muses on the way museums consecrate objects that were once meant for everyday use, but it also reveals that this protective approach may obscure significant information embedded in the objects: for example, the CT scans revealed a litany of damage, repairs, and excavations endured by an unassuming col-

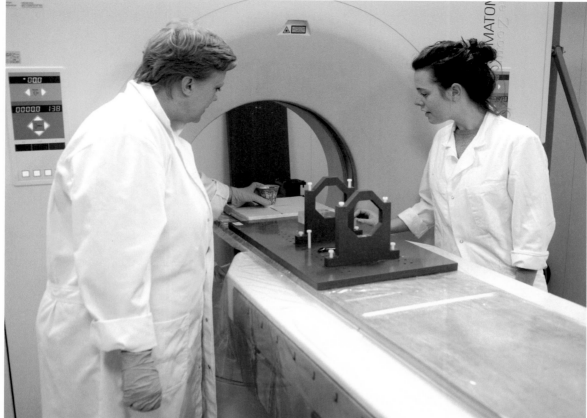

Analyzing the cup with a CT medical scanner

A seventeenth-century drinking glass (Museum Boijmans Van Beuningen)

A 3D print in plaster

lection of ceramic cups. Smart Replicas preserve the delicate formal beauty of these vessels, but connect them to contemporary technology using tools such as smartphone apps so that they are able to communicate a more complete narrative to museum visitors.

Smart Replicas on display at the Museum Boijmans Van Beuningen

Interactive software for tablet devices offers more information about artifacts

Center for Orthopedics Research and Development, Nemours/Alfred I. duPont Hospital for Children

Magic Arms – WREX

Year: 2013
Client/Purpose: For children with neuromuscular disabilities
Partners: N/A
Technique: Fused deposition modeling
Printer/Service: Stratasys Dimension
Material: ABS plastic
Production time: 1 day per piece
Budget: $2,000
Edition: On-demand

→ #body topology   → #building blocks   → #medical

Improving Mobility and Development for the Youngest Patients

tached to a custom-molded vest. It can also be modified in terms of size, range of motion, and resistance. Moreover, since the WREX can even be adapted for toddlers, it offers youngsters the prospect of improved development in terms of physical, cognitive, and emotional interaction with their environment, potentially granting them more freedom as they grow.

As the tools for medical analysis become more sensitive and precise, a deeper understanding of physical conditions allows an even more nuanced and individual diagnosis. The associated treatment, however, tends to operate in delay, favoring generic solutions given the high cost of customized prosthetics. At the Alfred I. DuPont Hospital for Children, an experiment in rapid prototyping is attempting to alter this disparity. Children with limited arm movements due to cerebral palsy, spinal cord injuries, muscular atrophy, multiple sclerosis, or other diseases are now being given the Wilmington Robotic Exoskeleton (WREX), a personalized device that combines 3D-printed parts, elastic straps, and metal brackets. It is an exoskeletal arm with two segments and four joints that can support fine motor operations and reduce fatigue. Because its parts are 3D-printed, the WREX is light and relatively easy to adapt for each patient: it can be mounted onto a wheelchair or at-

WREX exoskeleton made with ABS plastic

Richard van As

Robohand

→ #empowerment   → #body topology   → #open source   → #medical

An Open-Source Prosthetic for Unique Body Facilitation

Year: 2012
Client/Purpose: Research
Partners: MakerBot
Technique: Fused deposition modeling
Printer/Service: MakerBot Replicator 2
Material: PLA bioplastic
Production time: 3 months
Budget: $100,000
Edition: In development

In 2011, South African carpenter Richard Van As suffered an accident that required the amputation of four fingers from his right hand. When he researched his options for prosthetics, he quickly discovered the extraordinarily high costs associated with artificial limbs, in large part due to the need to produce each one as a unique piece, adapted to the wearer's body through laborious techniques. Van As achieved a provisional solution using his own wood workshop, but a much more significant opportunity eventually presented itself in a chance online encounter with American puppeteer Ivan Owen. Together, they began to work on a more sophisticated prosthetic with dexterous, flexible motions, but their first prototype, machined from aluminum, was already rather expensive and time-consuming to fabricate. Meanwhile, their work was attracting attention and requests from other people with missing digits. Van As and Owen turned to rapid prototyping, using a donated MakerBot 3D printer as an ideal way to work itera-

tively and produce affordable solutions, especially for children whose constant growth would repeatedly demand new fittings. Finally, they uploaded their designs for mechanical replacements of upper limbs and fingers to Thingiverse. As a result, anyone could download the plans for free and assemble the 16 printed parts and 28 off-the-shelf components for a total cost of about $150. A more recent version that employs snap-fit 3D-printed parts costs as little as $5.

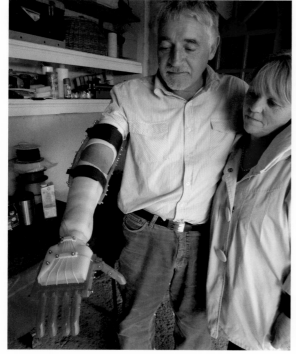

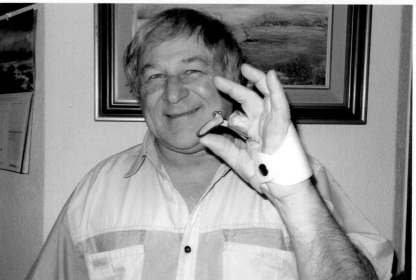

Jorge Lopes dos Santos
Fetus 3D Project

→ 3D scanning   → #body topology   → #intangible

Modeling Human Life as Sonic Geometries

Year: 2009 to present
Client/Purpose: Student research
Partners: MD Heron Werner, Ricardo Fontes, Feto3D, Royal College of Art
Technique: Polyjet, stereolithography, selective laser sintering, inkjet powder printing, fused deposition modeling
Printer/Service: Object350 Connex, Stratasys FDM Vantage, Stratasys uPrint, 3D Systems Spectrum Z510, 3D Systems Viper
Material: Resin, colored gypsum, ABS plastic
Production time: 48 hours per piece
Budget: N/A
Edition: N/A

When he first began the Fetus Project during his studies at the Royal College of Art in London, Brazilian designer Jorge Lopes dos Santos had no specific agenda beyond the form and technology. He simply began with an interest in the history of model-making, intimately entwined with the lineage of sculpture and the body; at the same time, the project has shared many links with medicine, from anatomical reproductions to medieval death masks. Lopes dos Santos eventually focused on

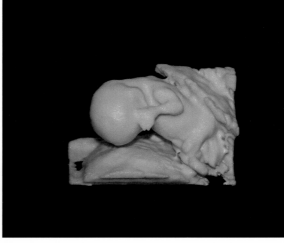

Ultrasound at 13 weeks

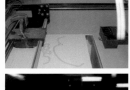

Printing model at 32 weeks

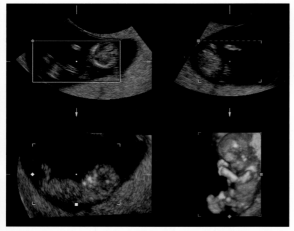

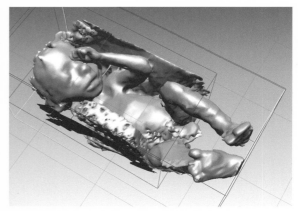

Virtual model of ultrasound at 18 weeks

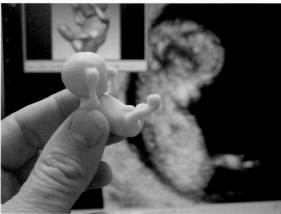

Ultrasound and printed model at 12 weeks

the challenge of 3D printing life-size models of babies in utero, including his own. Using a combination of ultrasound, computerized tomography (CT) scanning, and magnetic resonance imaging

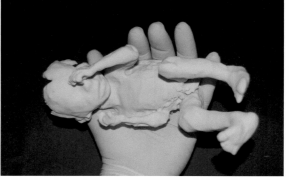

Printed model at 18 weeks

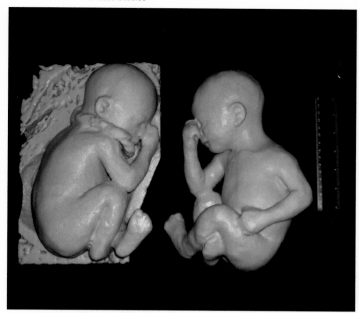

MRI scan at 32 weeks, with and without womb and umbilical cord

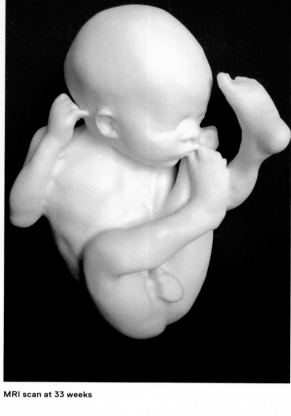

MRI scan at 33 weeks

(MRI), he developed a way to construct a three-dimensional model of the developing embryo and then to print it through rapid prototyping. The results are very powerful on an emotional level alone, but Lopes dos Santos's collaboration with the medical community revealed unexpected possibilities for the process. For example, it can be used in place of the conventional sonogram image for pregnant women who are blind; it can also help facilitate discussions with parents about deformed babies, and even aid doctors preparing for complex operations.

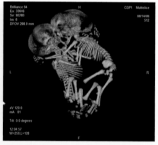

Skeleton of conjoined twins from CT

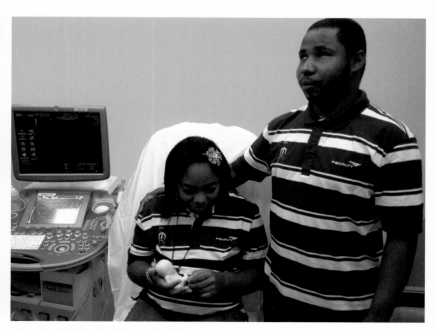

Blind parents holding a printed model of their future son

Year: 2010
Client/Purpose: For exhibition
Partners: N/A
Technique: Inkjet powder printing (human ash)
Printer/Service: 3D Systems ZPrinter
Material: Human ash
Production time: N/A
Budget: N/A
Edition: 1

Studio Wieki Somers
(Dylan van den Berg and Wieki Somers)
Consume or Conserve?

→ #materials

Reincarnating Human Life in Printed Furniture

While 3D-modeled objects can carry vast amounts of information beyond their simple form, the material in which they are printed tends to remain neutral and unquestioned (given the technical difficulty of innovating at this level of the process). Nevertheless, Wieki Somers imbues the very substance of a 3D-printed object with meaning in the "Consume or Conserve?" project. For the "In Progress" exhibition at Grand Hornu, Somers presented three still lifes printed in human ash as a meditation on the unexamined implications of consequences of exploiting materials in the natural environment to satisfy capricious desires. In this slightly macabre scenario, our very bodies might one day be subsumed into the voracious cycle of consumer consumption – and perhaps we might be discouraged from discarding the objects made of our own loved ones. In line with this mentality, Somers created a series of still lifes by pairing contemporary everyday objects with visual references from vanitas paintings of the Dutch Golden Age. These symbols – a scale and honeycomb, a

Birds and toaster, **still life**

technological progress without a clear ideological basis. The project practices an implicit form of self-critique, since it is itself a design innovation whose intention has remained intentionally vague and open. At the same time, "Consume or Conserve?" can be read as a subtle warning about the dung beetle and vacuum cleaner, and a bird and toaster – evoke an existential attitude toward human life, as brief and futile as death is certain.

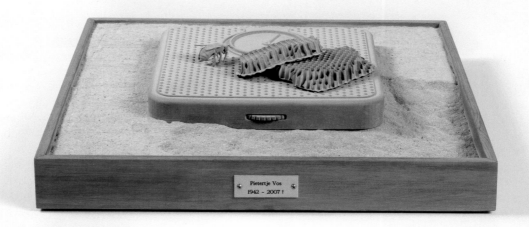

The weight of a honeycomb, **still life**

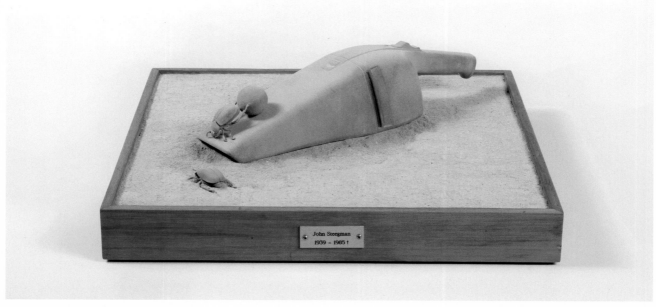

Dung beetle and hand vacuum cleaner, **still life**

Maya Ben David
Save As [   ]

→ #intangible

Embedding Objects with Collective Experience

Year: 2011
Client/Purpose: Student research
Partners: Stratasys, Design Academy Eindhoven
Technique: Polyjet
Printer/Service: Objet Connex
Material: VeroWhite acrylic photopolymer
Production time: 5 months
Budget: $70 – 350 per print
Edition: 1

Capturing

Editing

Designing

Producing

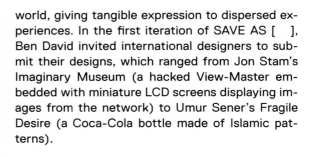

Since the rise of the blog web format in the late 1990s, individuals have developed increasingly sophisticated ways of confronting the massive amounts of news and information that we consume: today, we have multiple formats, such as Tumblr, for the interactive digital "curation" of content. Nevertheless, our means for reacting to this information and representing it in other ways remain stubbornly virtual. Although craft skills are experiencing a revival with the dissemination of the tools and knowledge needed for self-production, the rapid news cycle makes permanent responses untenable – which may explain our propensity to quickly forget what we have observed. Israeli designer Maya Ben David has proposed a new way of sharing and archiving collective memory with her project SAVE AS [   ]. The online platform enables the collection of text and images in relation to globally significant events, such as the Arab Spring of early 2011. It also supports contributions of physically encoded reactions in the form of uploaded virtual models. These models can then be downloaded and 3D-printed by anyone around the world, giving tangible expression to dispersed experiences. In the first iteration of SAVE AS [   ], Ben David invited international designers to submit their designs, which ranged from Jon Stam's Imaginary Museum (a hacked View-Master embedded with miniature LCD screens displaying images from the network) to Umur Sener's Fragile Desire (a Coca-Cola bottle made of Islamic patterns).

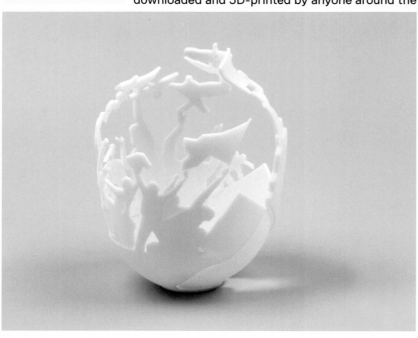

SAVE AS ['Revolution Vase']

SAVE AS ['Fragile Desire']

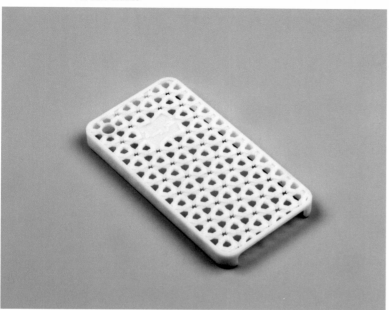

SAVE AS ['Freedom']

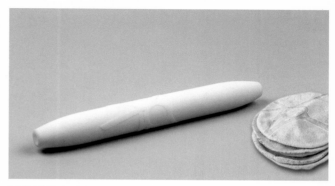

SAVE AS ['Tahrir Rolling']

SAVE AS ['Imaginary Museum']

Lorna Barnshaw
Replicants

Year: 2013
Client/Purpose: Artwork
Partners: N/A
Technique: Inkjet powder printing
Printer/Service: 3D Systems ZPrinter 650
Material: Colored gypsum
Production time: 4 months
Budget: $150
Edition: 1

→ #3D scanning   → #body topology

The Uncanny Valley of the Scanned Body

Like digital animation, 3D printing can be extremely impressive when it takes on an abstract or cartoonish aesthetic of bright colors and smooth geometric forms. When it is used to manifest realistic objects, especially anthropomorphic ones, the results can be altogether unsettling – as in the "uncanny valley" of very realistic human animations, which can have a slightly dead or zombie-like quality. With Replicants, Lorna Barnshaw explored the possibility of reproducing human faces using

a full-color 3D printer that binds sandstone powder with resin and colored ink, much like a standard inkjet printer. Each of the three facial fragments, like the ruins of ancient sculptures, represents a different process of image capture and photo stitching for three-dimensional reconstruction. Autodesk's 123D Catch created a painterly but incomplete surface from 40 photos, while Cubify Capture extracted a very pixelated and distorted facial image from a video file. The VIUscan 3D scanner generated the most lifelike (and also the most unnerving) result: the face resembles a preserved death mask with glitches and artifacts from the scanning process that suggest the onset of decomposition. Replicants is disconcerting to the extent that its underlying technology is imperfect, yet the increasing sophistication of this technology may yield something even more alarming – a believable simulacrum of a living being, more concrete than reality itself.

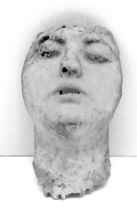

Facial reconstructions made with three different modeling processes

Heather Dewy-Hagborg
Stranger Visions

→ #intangible

Facial Reconstructions from Genetic Traces in Public Space

Year: 2013
Client/Purpose: Artwork
Partners: Genspace, Clocktower Gallery, Eyebeam
Technique: Inkjet powder printing
Printer/Service: 3D Systems ZPrinter
Material: Colored gypsum
Production time: N/A
Budget: N/A
Edition: 1

Compared to analog information, text, or images, digital content is easier to transfer and duplicate with minimal loss in quality. It can also be more rapidly encrypted or analyzed, and more efficiently stored in larger quantities. Nevertheless, the public consciousness remains more sensitive to physical manifestations of data than to the data itself, as demonstrated by reactions to the 3D-printed gun. The Stranger Visions project by Heather though Stranger Visions is positioned as an artistic provocation, it is also a harbinger of society's ineluctable confrontation with progressively decentralized technologies.

Traces of DNA collected from public space

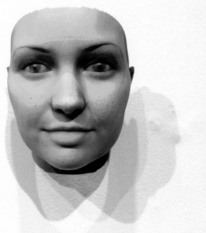

Dewey-Hagborg has sparked a similar discussion about genetic data. Although the collection, testing, and interpretation of DNA is widely accepted for the purposes of identifying criminals and tracing diseases, Stranger Visions is shocking in its material appropriation of our very identities. Dewey-Hagborg collected discarded objects from the streets of Brooklyn, such as cigarettes and gum, and submitted them to Genspace, a DIY biology lab, to map the DNA found in traces of saliva. She then isolated single nucleotide polymorphisms, the specific genetic chains that determine variables such as hair, skin, and eye color,mand body and facial morphology. By feeding these variables into custom software, she could model faces that reflected the characteristics of each sample. Finally, the results were 3D-printed in full color and displayed in New York's Clocktower Gallery. Al-

Stranger Visions installed at Clocktower Gallery, New York

Laureline Galliot
LINE & MASS

→ #interact

From Wireframe to Painted Volume

Year: 2012
Client/Purpose: Student research
Partners: École nationale supérieure de création industrielle
Technique: Inkjet Powder Printing
Printer/Service: 3D Systems ZPrinter 650
Material: Colored gypsum
Production time: 6 months
Budget: N/A
Edition: 1

Line & Mass **from left to right:** *Teapot, Tool 3Dpalette,* **Mask, and** *Self-Portrait* **(oil on canvas)**

Mask inspiration

Line & Mass **research. Sugru and Plasticine mock-up**

The physical act of 3D printing has received an extreme amount of attention in the public consciousness, while the other half of the process – virtual modeling – is usually taken for granted. Unfortunately, no matter how much progress and experimentation takes place in the hardware and material output of 3D printers, a lack of innovation on the virtual end may be responsible for the limited conception of what rapid prototyping can achieve. LINE & MASS, the diploma project of Laureline Galliot at ENSCI in Paris, suggests that the normal way of modeling in pristine wireframes is not the only option. In developing a new option, she drew on a variety of inputs, from her own paintings and collection of objects to the nineteenth-century argument between Ingres and Delacroix on the respective merits of line and mass. She also incorporated her experience working with Disney on tactile animations. All of this was assimilated into an original method of digital drawing that revolved not around contours but aggregations of solid, textured color based on a personal palette. By appropriating intuitive modeling software ostensibly made for digital animators, such as Zbrush, Galliot could paint, sculpt, and freely rotate the virtual object. It was also possible to sample digital images as texture maps, whereas design software emphasizes orthogonal views and geometric precision. The LINE & MASS collection, consisting of a "tool", a teapot, and a mask, serve to illustrate the possibilities of this new approach.

Digital paintings

Line & Mass **Teapot**

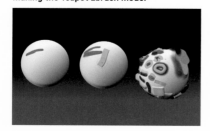

Making the Teapot zbrush model

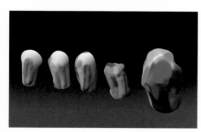

Making the Tool 3Dpalette zbrush model

Tool 3Dpalette, Mask, and Teapot

Gradient Bangles exist as both digital and wearable art

Maiko Gubler
Gradient Bangles

→ #wearables

Hybridizing Digital Art and Wearable Jewelry

Year: 2013
Client/Purpose: Artwork
Partners: N/A
Technique: Inkjet powder printing
Printer/Service: 3D Systems ZPrinter
Material: Colored gypsum
Production time: N/A
Budget: N/A
Edition: 25

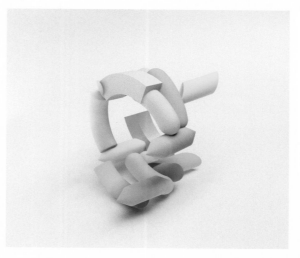

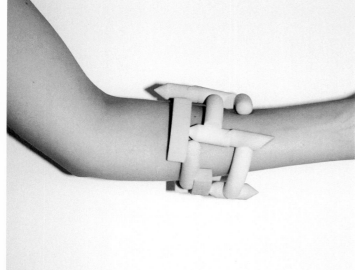

The work of Berlin-based artist Maiko Gubler treads an intentionally precarious path between different fields and media; the true "identities" of her objects, rendered in multiple formats from digital image to printed object, are difficult to locate, especially when the issue of material translation arises as a concretization (or, on the contrary, as a compromise) of the pure virtual creation. Her series of Gradient Bangles, for example, began as computer renderings, their platonic geometries then being warped into wearable bracelet form, and the surfaces developed as pastel gradients of color based on the mathematical calculations of surface-normal mapping. In particular, the ∞Limited version is further generated through self-forming topology, with an infinite number of possible permutations. The virtual designs are then 3D-printed in full color with a gypsum powder substrate that gives the bangles a chalky characteristic. Gubler's work is part of a larger movement within full-color printing that avoids photorealism to explore the affordances native to the technique, where the exact manifestation of a digital form creates a nebulous artifact that is neither completely real nor unreal.

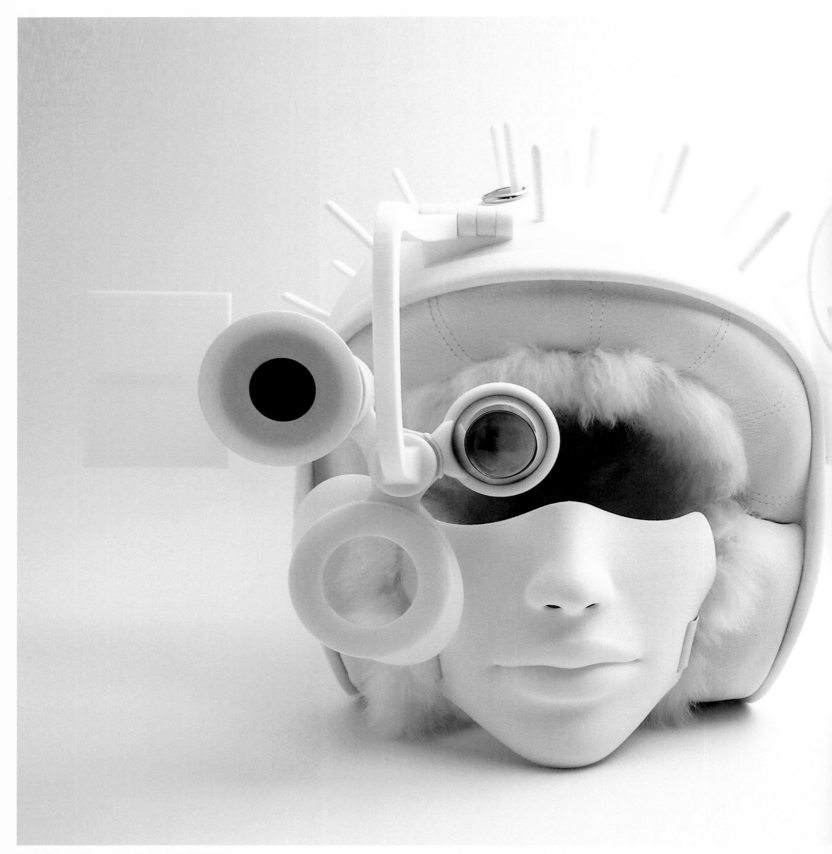

Trophy Helmet (**Signature: gold-sheet ATN 1/3 2012**)

Atelier Ted Noten
7 Necessities

→ #wearables

The Modern Woman's Survival Kit Blends Nylon with Jewels and Fur

Year: 2012
Client/Purpose: For exhibition
Partners: Organisation in Design
Technique: Selective laser sintering
Printer/Service: Materialise, Freedom of Creation
Material: Nylon, glass, crystal, acrylic, gold, silver, diamond, fur
Production time: 1 year
Budget: N/A
Edition: 3

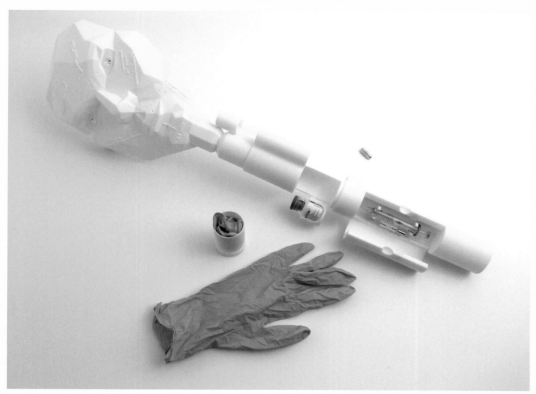

Beauty Mask **(backside silver oval ATN 1/3 2012)**

For contemporary designers, jewelry is an extremely open field characterized more by its specialized techniques than by the function of the resulting objects. Ted Noten uses this freedom to postulate on the self-indulgent, macabre elements of modern-day luxury, but his methods tend to originate in traditional crafts such as goldsmithing, stonesetting, and even taxidermy. He began to explore 3D printing in part to question the relevance of these laborious crafts and in part to test the limits of rapid prototyping as an expressive medium. His 7 Necessities collection can thus be seen both as artifacts of a hybrid material language juxtaposing plastic with diamonds and as cultural devices that reveal the self-optimization of society through technological means. Noten's "survival kit for the modern woman" includes a Botox-injecting beauty mask that sculpts the wearer's face into the classical paradigm of Nefertiti (accompanied by a gold suicide pill in case of failure); a Dior-branded gun fitted with hidden compartments holding lip-gloss, a USB stick, and diamonds; a fur-lined helmet equipped with radar devices, a telescope for women "on the hunt," complete with spikes to display the rings she collects as spoils; and other "essential" objects. Working with a 3D designer to fine-tune the results, Noten elevates 3D printing to an artistic caliber by treating it as the interface between age-old material connotations and a rapidly evolving cultural context.

Purse of wonders (**Signature: printed in nylon ATN 2011 Nr (7) 0001**)

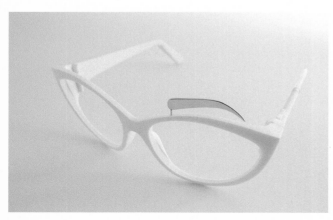

Glasses for a women on the warpath (**14 kt gold sheet ATN 2/7 2012**)

Chatelaine (**14 kt gold sheet ATN 1/3 2012**)

Chastity Belt **(14 kt gold sheet ATN 1/3 2012tion)**

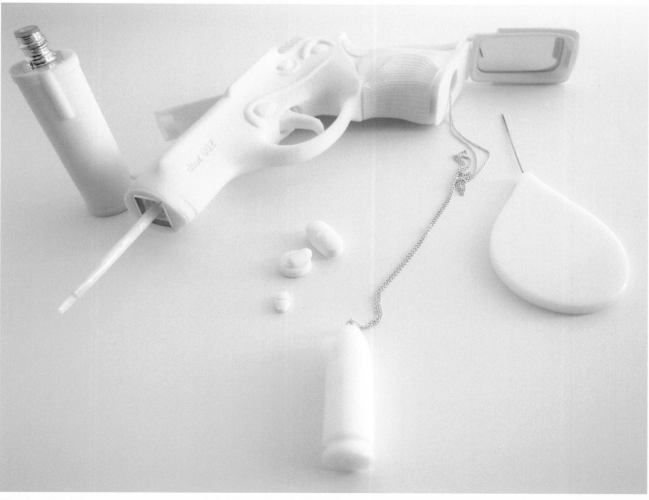

White Gun **(Signature printed in the gun: Dior 002 ATN20121510 (date of fabrication))**

Atelier Ted Noten
Haunted by 36 Women

Year: 2009
Client/Purpose: For exhibition
Partners: Organisation in Design
Technique: Selective laser sintering
Printer/Service: Freedom of Creation
Material: Nylon
Production time: 1 year
Budget: $100 per print
Edition: N/A

→ #wearables

Jewelry from Scaled-Down Sculptures of Female Archetypes

Miss Piggy **ring, part of Haunted by 36 Women**

Dutch jewelry designer Ted Noten became famous for his iconic acrylic handbags, fusing the predatory attraction exerted by luxury items with the contradictory desires to display and conceal wealth, and to indulge in and suppress vice. By casting forbidden items like guns or cocaine in solid resin, he presented a conflicted modern female identity in "freeze-frame" capture. Eventually, Noten began to broaden his signature language with forays into 3D printing, although his method differs significantly from the conventional routine of modeling directly in a virtual environment. In the Haunted by 36 Women series, for example, he created a series of sculptures based on a variety of archetypal characters, from the femme fatale and the nymphomaniac to the suffragette and the girl-next-door, using an eclectic range of materials, including rubber tires, stiletto shoes, boxing gloves, machine guns, chairs, and stuffed animals. Working with a 3D modeler, Noten scanned these assemblages and scaled them down digitally before refining the resulting forms through an iterative rapid prototyping process. The final outcome was a collection of rings, necklaces, and brooches, made wearable by the miniaturization of the source material. While producing one of each archetype in gold, he also released the models in small batches of laser-sintered nylon prints.

Ice Cream Girl, **part of Haunted by 36 Women**

Pig Bracelet, **part of Haunted by 36 Women**

Nervous System
Rhizome Cuff

→ #customization   → #wearables

Unique Jewelry Generated as a Non-Linear Structure

The rhizome, a complex stem structure that grows horizontally, capable of sprouting roots or shoots from any of its multiple nodes, emerged as a philosophical trope in *A Thousand Pleateaus*, written in 1980 by Gilles Deleuze and Félix Guattari. Their research focused on the non-linearity, indeterminacy, and spontaneous resolutions and dissolutions that occur in rhizomatic structures in contrast to the unidirectional relationships embedded within tree structures. Over the next few decades, this idea would become one of the most significant operational modes for architecture and design, which departed from the traditional hierarchical creative approaches in search of complexity and rule-based simulations. While most rhizomatic designs remained confined to paper and computer screens, Nervous System has used additive manufacturing to physically manufacture objects in this vein. The eponymous Rhizome Cuff, part of the studio's Hyphae collection, is realized through generative scripts that mimic the complex growth patterns of natural systems, merging and branching in unique ways at each point in the structure. The cuff is then printed in laser-sintered nylon or cast in silver or brass using the lost-wax technique.

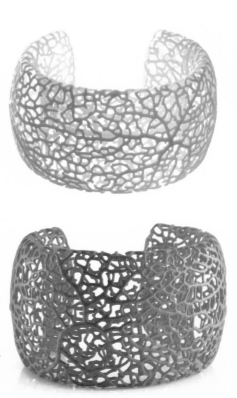

Iris van Herpen
VOLTAGE

→ #materials   → #wearables   → #body topology

Complex Printed Clothing Based on Biological Skins

Year: 2013
Client/Purpose: For fashion show
Partners: Stratasys, Neri Oxman
Technique: Polyjet
Printer/Service: Objet500 Connex
Material: Varied
Production time: N/A
Budget: N/A
Edition: 1

Iris van Herpen is the youngest member of the official Chambre Syndicale of Haute Couture in France and has become strongly associated with high-technology fashion, especially the use of 3D printing. In fact, her work employs many techniques in addition to additive manufacturing, such as hand-finishing, laser-cutting, and digital knitting. The combination of these methods varies from collection to collection, but all of van Herpen's items are united in her evolving search for fractal complexity. The 2013 catwalk show for her Voltage collection featured her most sophisticated use of 3D printing to date, including a cape and skirt designed in collaboration with architect and designer Neri Oxman from the MIT Media Lab. Drawing on van Herpen's experience in shaping complex systems to the human figure and Oxman's experiments with rapid-prototyped, biomimetic structures in her Imaginary Beings collection, the outfit was made with Stratasys using advanced multi-material printing: an armor-like shell of thousands of hard conical spines moves flexibly over the surface of the skin, like an animal pelt, while its shape references traditional haute couture, which tends to sit away from the body to display the designer's technical and sculptural skills. While van Herpen's career path moves from the edges of the avant-garde to the more mainstream apparatus of the Parisian fashion world, she also maintains her openness to unexpected partnerships outside of its normative boundaries.

Iris van Herpen × Rem D. Koolhaas
Wilderness Embodied

→ #wearables   → #complexity

Wrapping the Foot in a Tangle of Roots and Tendrils

Year: 2013
Client/Purpose: For fashion show
Partners: Stratasys
Technique: Polyjet
Printer/Service: Objet Connex, Objet Eden
Material: VeroBlackPlus, VeroWhitePlus acrylic p hotopolymer
Production time: N/A
Budget: N/A
Edition: N/A

As the founder of the shoe company United Nude, Rem D. Koolhaas was one of the first to apply a conceptual design mentality, in his case acquired from an architectural background, to the highly specialized craft of shoemaking. He designed footwear that eschewed the conventional tropes of the fashion world in favor of more eclectic themes, such as the Lo Res faceted high-heeled rubber shoe, created using the vocabulary of virtual modeling. He and Iris van Herpen shared a natural affinity that led to collaborations such as the intricately webbed Crystallization design and the taloned Fang shoes made from molded fiberglass. Nevertheless, the complexity of the 3D-printed shoes they designed for van Herpen's first haute couture show in Paris represent an unprecedented level of complexity in achieving a symbiosis between wearability (something United Nude is renowned for, despite their unorthodox approach) and form. Produced by Stratasys on extremely high-resolution machines, the form is only made possible by 3D printing: the open tangle of root-like tendrils could not be made using any other method of production. Rather than revealing the constraints of the medium, the shoes point to unexplored territory in a field that is already rife with innovation.

Marloes ten Bhömer
Rapidprototypedshoe

Year: 2010
Client/Purpose: For exhibition
Partners: Objet
Technique: Polyjet, selective laser sintering
Printer/Service: Objet500 Connex
Material: VeroBlack acrylic photopolymer, TangoBlack photoelastomer
Production time: 6 months
Budget: N/A
Edition: 1

→ #wearables → #materials → #body topology

Specialist Craft Reduced to a Single Material Process

More than a decade ago, Marloes ten Bhömer made her first investigations into 3D-printed shoes, a territory that she largely staked out as her own in the years that followed. Accustomed to using unconventional materials in her designs, she further circumvented the strict boundaries traditionally found in the highly technical craft of shoe-making (as well as the reluctance of specialists in other industries to make parts for her unusual designs) with her design for SLSshoe. The nearly indiscriminate nature of 3D printing made the process rather effortless, though it was not optimized for comfortable results. Seven years later, with Rapidprototypedshoe, ten Bhömer adapted her design approach to the possibilities and limitations offered by additive manufacturing in order to produce a more viable shoe. She quickly realized that the wear and tear incurred by footwear make certain details and articulation of parts desirable, even if they are unnecessary from a manufacturing perspective. Therefore, she decided to create a shoe that could be printed in one pass but also disassembled for later repair and replacement. At the same time, the shoe was engineered at the microscopic level to combine two different kinds of plastic, one rigid and one flexible, in varying densities for a more comfortable result. Ultimately, ten Bhömer's shoe contends with the same problems as any shoe – support, comfort, and freedom of movement – but its distinctive aesthetic reflects an entirely different manufacturing technique.

Jiri Evenhuis
Paris

→ #wearables

The Micro-Architecture of Footwear

Year: 2004
Client/Purpose: N/A
Partners: N/A
Technique: Selective laser sintering
Printer/Service: Freedom of Creation
Material: Polyamide
Production time: N/A
Budget: N/A
Edition: N/A

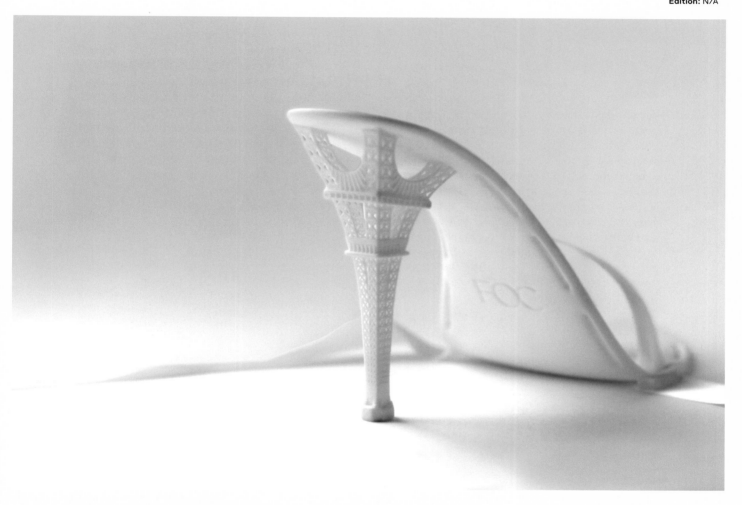

The current developments in additive manufacturing tend toward the increased complexity of multi-material printing in order to achieve more sophisticated and diverse functions. In light of the industry's extremely rapid pace of technological adaptation, experiments from only ten years ago can seem almost charmingly naïve. However, it is worth reflecting on the context in which objects like Jiri Evenhuis's Paris stiletto appeared: presented at the Dyson showroom during the Paris Design Festival in 2004, the technique did not yet dare venture into the realm of fashion (although it has since made a confident arrival). The shoe is almost poetically minimal in its construction, save an inverted Eiffel Tower for its heel. Today, such simplicity could be seen as a drawback considering the functional challenges posed by the design of footwear; at the time, though, the object expressed a prescient approach to craft, translating the extremely specialized skills of shoemaking into the digital language of code, adaptable to the extremes of pastiche, freely-sampled references, and fine-tuned, personalized scans of the human body's complex surfaces.

byAMT Inc

Jointed Jewels
(Collections 2008 – 2009, 2010, and 2011)

→ #assembly free → #customization → #wearables

The Industrial Ball Joint Transformed into Decorative Jewelry

Year: 2008 to 2014
Client/Purpose: For sale and exhibition
Partners: Uformia
Technique: Selective laser sintering
Printer/Service: N/A
Material: Coated nylon
Production time: 3 to 6 months per collection
Budget: $25 to $1,130 per print
Edition: N/A

Selective laser sintering is uniquely suited to the production of complex forms: as a laser draws a path through a bed of tiny particles, the desired shape is fused together in one pass combining any nested or interlocking parts, and the excess particles act as a supporting structure without the need for additional struts. In Jointed Jewels, Alissia Melka-Teichroew of byAMT Inc. has used the process to make accessories inspired by the ball joint, a technical component often found in more industrial applications. The project began in 2008, and the essential schematic of the ball joint of the jewelry has since been interpreted as a "family" of objects in various iterations, including experi-

ments with quilted textures and coloring (industrial and hand-dyed). The latest development, created in partnership with Norwegian company Uformia, introduces the possibility of user customization via an online interface. Using Uformia's Symvol software, which mimics the real world of solid objects rather than the virtual language of infinitely thin surfaces, Melka-Teichroew is able to explore new levels of physical detail and grain in generative modeling through a graphic interface in the Rhino software.

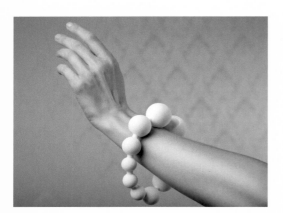

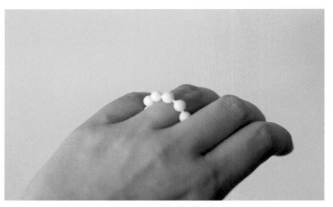

The interlocking ball joints are printed in a single pass

Studio Femke Roefs
3D-Printed Buttons

→ #wearables

Wearable Ornaments Drawn in Plastic Spirals

Year: 2012
Client/Purpose: Research
Partners: Leoni Werle
Technique: Fused deposition modeling
Printer/Service: RepRap Mendel
Material: PLA bioplastic
Production time: 6 months
Budget: $1 per print
Edition: In development

3D printing is generally used to achieve continuous planes modeled in virtual environments, as if one were casting the surface of the model (like in older industrial techniques), albeit with a grainier, striated surface. Of course, additive manufactur-

"drawing" quality of 3D printing and wanted to exploit, rather than hide, this quality. By adopting the language of the spirograph, a children's drawing tool launched in 1965, they were able to design lacy, more open structures in circular patterns,

ing has its own advantages over traditional casting, such as the elaborate internal support structures that can often be used for lighter, stronger visual results by tracing each cross section. Designers Femke Roefs from Eindhoven and Leoni Werle from Cologne were both fascinated by the

using the line patterns as a design feature rather than viewing them as a drawback. The resulting creations can be worn as buttons, establishing a link between a more traditional craft and a contemporary technology.

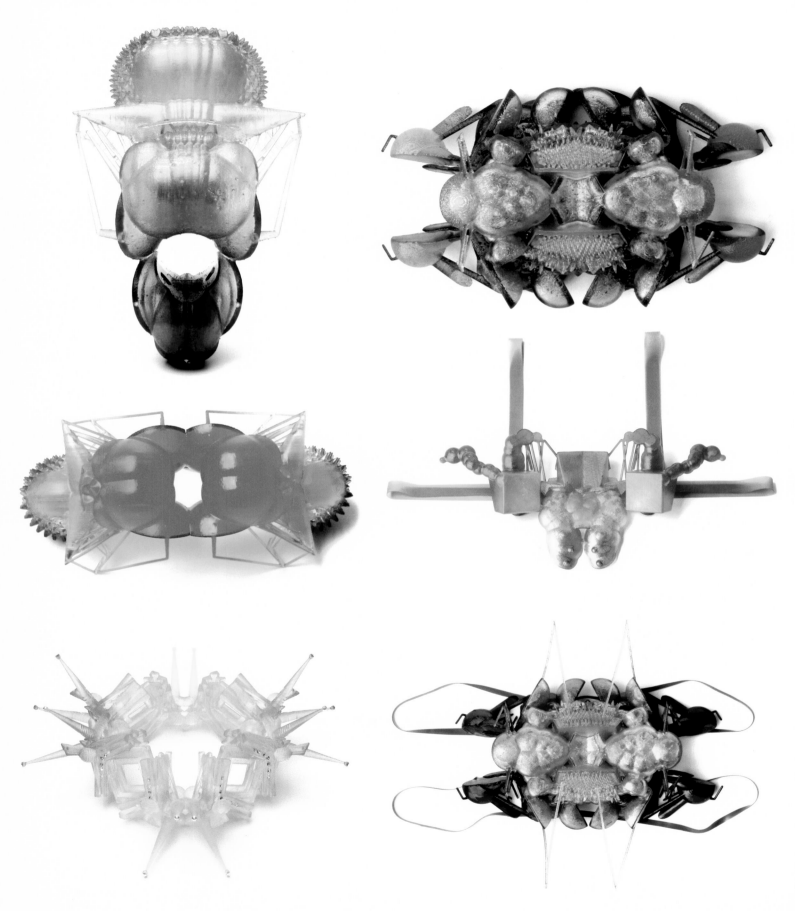

Dorry Hsu
The Aesthetic of Fears

→ #wearables → #intangible

Insectile Jewels Drawn with Digital Intuition

Year: 2013
Client/Purpose: Student research
Partners: Royal College of Art
Technique: Stereolithography
Printer/Service: N/A
Material: Resin, latex, photopolymer
Production time: 2 months
Budget: $1,300 total/ $410 per piece
Edition: 1

When discussing the aesthetics of 3D printing, many people conflate the issue with the aesthetics of 3D modeling; in fact, while the former can be rather grainy and full of small imperfections, the latter is abstract, scaleless, and as perfect (or flawed) as the modeler likes. Yet, for many digital designers, the practice of 3D modeling itself still tends to favor certain qualities (such as repetition, symmetry, and planarity) over more variable outcomes. In the Aesthetic of Fears project, however, designer Dorry Hsu achieved a unique visual language for her 3D-printed insectile jewels, in part because they were created using the Geomatic Freeform software and a "haptic arm." By hold-

ing a stylus, Hsu could push, pull, and twist her creations as if she was forming them from clay, receiving physical feedback from the resistance of the stylus to her motions. Hsu later elaborated the forms with more detailed modeling, and finally printed them in transparent resin and coated them by hand with colored latex. In line with her speculative approach to the future use of jewelry, Hsu's process takes the familiar appearance of additive manufacturing and makes it beguilingly unfamiliar and subtly nuanced.

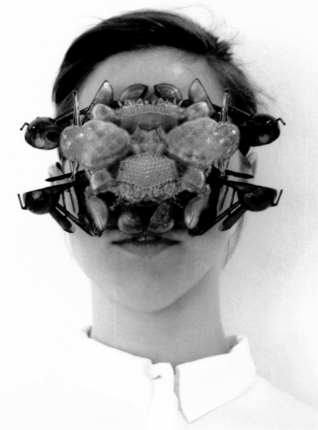

Janne Kyttanen
Lost Luggage

Year: 2000 to 2014
Client/Purpose: Research
Partners: 3D Systems
Technique: Selective laser sintering
Printer/Service: 3D Systems
Material: Nylon
Production time: N/A
Budget: N/A
Edition: 10

→ #assembly free   → #wearables

Reprinting your Belongings on Arrival

More than any other method of production, the value of the object in digitally controlled manufacturing techniques lies in the virtual blueprint rather than the physical print, simialr to the difference between an online inbox and a physical mailbox. However, 3D printing remains mostly framed as an intervention into fixed physical locations, such as the home, the laboratory, or the factory. Janne Kyttanen, the co-founder of design company Freedom of Creation, imagines a different scenario: what if 3D printing could make luggage obsolete? Could we travel unencumbered and recreate the necessary ob-jects on-site? Since 2000, Kyttanen has developed a variety of larger-scale objects, including clothes, shoes, gloves, a hat, a scarf, and a large attaché bag, broadening the scope of what we might expect from additive manufacturing on an everyday basis. Perhaps, then, our increasingly mobile lifestyles will no longer require us to carry these things from place to place; instead, we would each own a unique digital collection, tailored to our size, taste, and needs, and we would access the necessary print data at each new location, just like we can log in to check our e-mail from any computer in the world that has an Internet connection.

Elvis Pompilio

H.MGX, Spirograph.MGX, and Lys.MGX

Year: 2012
Client/Purpose: For sale
Partners: .MGX by Materialise, .rad Product
Technique: Selective laser sintering
Printer/Service: Materialise
Material: Polyamide
Production time: N/A
Budget: $280 – 675
Edition: On demand

→ #complexity → #wearables

Bespoke Millinery Produced on Demand

When hats were obligatory in everyday dressing for public life, the structural demands on their construction required extensive knowhow as well as professional equipment such as hat blocks, steamers, and specialized textiles to make a soft, flattering shape that would retain its form despite daily handling and weathering. However, with most hats now being a fashion accessory for occasional wear, these challenges have been superseded by the opportunity for extravagant visual gestures and unusual materials. Like other Belgian designers emerging in the 1980s, such as the Antwerp Six and Martin Margiela, designer and milliner Elvis Pompilio is influenced, but not constrained, by historical technique: his work ranges from a cowboy hat for Madonna to a wireframe top hat and other bespoke marvels. In 2012 he began collaborating with Materialise to create a hat that would

be every bit as imaginative as previous creations, but also available for consumer purchase. Spirograph.MGX, designed with Igor Knezevic, is a swirl of thin ribbons that can be worn wherever the user desires, while Lys.MGX is a filigree veil that cradles the head and face. Pompilio later reinterpreted the fedora in H.MGX, making the dome and brim from a lacy "fabric" of interlocking fleur-de-lis motifs, connected with a flexible chain-link band. Similar to the traditional fedora, the hat is molded to the shape of the head, but for the purpose of 3D printing, this complex curvature is achieved using sophisticated digital modeling.

Elvis Pompilio × Igor Knezevic

Lys.MGX

Axel Brechensbauer
Bird Eyes

→ #wearables

Digitally Challenged Nature, Finished by Hand

Year: 2012
Client/Purpose: Artwork
Partners: N/A
Technique: Selective laser sintering
Printer/Service: Shapeways
Material: Nylon
Production time: N/A
Budget: $475 per print
Edition: 1

Axel Brechensbauer's Bird Eyes are a curious object – retro in their appropriation of the popular Ray-Ban Wayfarer sunglasses, technologically advanced in their fusion of sampled forms with simple geometric volumes, and contemporary in their hand-finished texture. The Swedish artist, who is based in Barcelona, produced this piece using the DAVID Laserscanner software, a low-cost solution for 3D scanning that uses a digital camera or webcam, a handheld laser, and two boards. He then remodeled the sunglasses digitally and had them printed in laser-sintered nylon via Shapeways before sanding them by hand and spray-painting them black to create the final appearance. Bird Eyes follow several of Brechensbauer's axioms ("Aim to be in the conflict zone between shapes." "Flat surfaces are essential." "Don't mimic nature – challenge it." and "Nature is ugly as it is; nature has to be forced to beauty."). These notions, assisted by the potential of contemporary technology, are behind the appearance of this artistic artifact.

Catherine Wales
Project DNA

Year: 2012
Client/Purpose: Student research
Partners: London College of Fashion
Technique: Selective laser sintering
Printer/Service: EOS Formiga P110
Material: Nylon
Production time: 8 – 20 hours per print
Budget: $300 – 8000 per piece
Edition: Unique piece

→ #building blocks → #3D scanning → #wearables →#body topology

Customizable Haute Couture from Full-Body Scans

Fashion has resisted the influence of 3D printing longer than other fields for a variety of reasons, such as the challenge of scale, the need for flexibility and comfort, the mismatch in appropriate materials, and the challenges of the fashion market, which still employs a huge number of manual workers across the world. Where clothing has been 3D-printed, it has been almost exclusively in the context of haute couture, one-off creations. sizes. Rather than using the normal scaling conventions for pattern cutting, Project DNA works like a kit of parts based on a digital avatar, which can be generated by scanning an individual's body. A personalized outfit can then be built up from a combination of a corset, breastplate, shoulder pieces, and hip panniers built like a space frame, and further adapted with moveable feathers that attach to the underlying structure via custom

With Project DNA, meanwhile, Catherine Wales has approached the fusion of the two fields from another perspective. While completing her master's degree at London College of Fashion, Wales began to speculate about the possibility of garments without size labels, taking into account the specific shape of each woman's body and avoiding the psychological issues associated with clothing joints. Working with London-based Digits2Widgets, Wales was also able to refine the thickness of the nylon feathers to create a softer effect.

Catherine Wales
Project DNA

Moto Waganari
Real Virtuality

→ #complexity

The Human Form in Multiple Dimensions

Year: 2009 to present
Client/Purpose: Artwork
Partners: N/A
Technique: Fused deposition modeling, selective laser sintering
Printer/Service: Shapeways, Alphaform
Material: PLA bioplastic, polyamide
Production time: 1 month per piece
Budget: N/A
Edition: 3–8

"Welcome to Borderland. Welcome to a place where reality is seduced by virtuality." So begins the experience of entering the online gallery of Moto Waganari, a sculptor who works between two and three dimensions in both digital and physical contexts. Waganari's figural sculptures – seated, falling through space, augmented with rabbit ears, and cropped into various sections – are born as virtual models and uploaded to an online archive where they can be rotated, manipulated, and even geometrically distorted by the viewer. At the same time, they are also manifested as 3D-printed wireframes and presented in physical space, posed with dramatic lighting to reveal their multiple profiles and dimensions with stark shadows being cast on the gallery walls. The visual language employed in these works is strongly derived from a relatively recent aesthetic of 3D models, from pure Platonic solids to complex tessellations. As the surface of the sculptures becomes more detailed and finely faceted, it approaches a palpable solidity with increasingly anthropomorphic characteristics and individual "personalities." Given the directness of rapid prototyping, these pieces maintain a dual existence both as objects and as simulations.

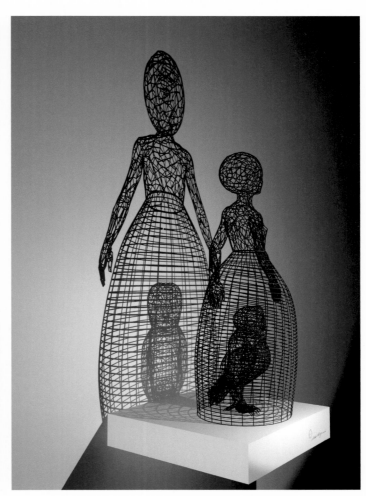

Studio Francis Bitonti
Dita's Gown

Year: 2013
Client/Purpose: Dita von Teese
Partners: Michael Schmidt, Shapeways
Technique: Selective laser sintering
Printer/Service: EOS Formiga P100
Material: Nylon, hematite crystal
Production time: 4 months
Budget: N/A
Edition: 1

→ #wearables → #complexity → #body topology

Draping Articulated Fabrics on the Digitized Body

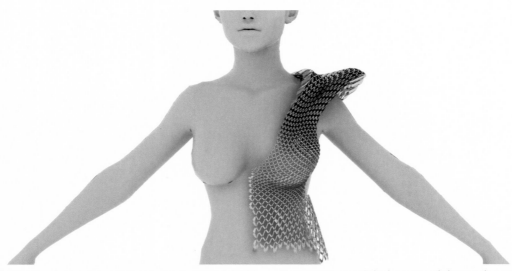

Historically, bespoke clothing has always contained an element of customization, albeit one achieved through a laborious and incredibly expensive process of hand-tailoring clothing samples to the individual customer's body. While the 3D-printed dress made for burlesque performer Dita von Teese is hardly an example of affordable design (after all, it is covered in 12,000 Swarovski crystals), it does demonstrate a complete transformation of the manufacturing process. Emerging from a collaboration between Francis Bitonti Studio, Michael Schmidt, and Shapeways, the dress was modeled as an elaborately articulated, flexible surface over Dita's scanned body, and printed using laser-sintered nylon in large sections that were later dyed and assembled by hand. More significantly, the dress eschews the primary techniques of clothing construction (needle and thread, padding, bonded fabrics, and so on) and adopts a language closer to that of product design or architecture, reflecting its creators' backgrounds. At the same time, the skillful modeling introduces qualities that are still rare in 3D-printed objects, such as fluidity and draping.

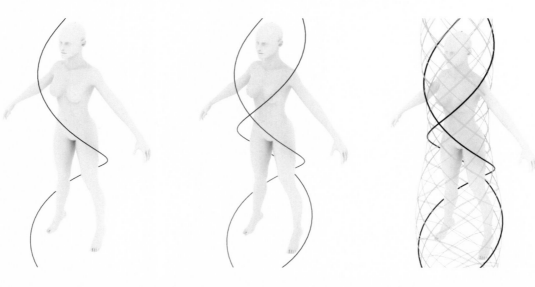

Nervous System
Kinematics

Year: 2013
Client/Purpose: For sale
Partners: N/A
Technique: Selective laser sintering
Printer/Service: Shapeways
Material: Nylon
Production time: 6 months
Budget: $25 to $350 per print
Edition: On demand

→ #new industries → #complexity → #customization

Articulated Geometries in Flexible Chain-Links

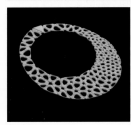

Necklace

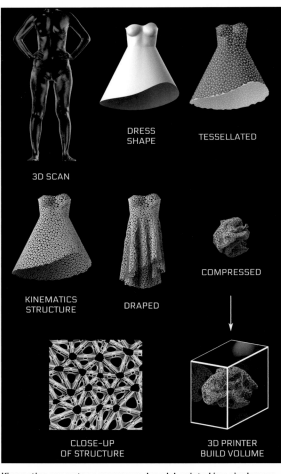

Kinematics generates a compressed model, printed in a single pass, that unfolds to become a flexible dress

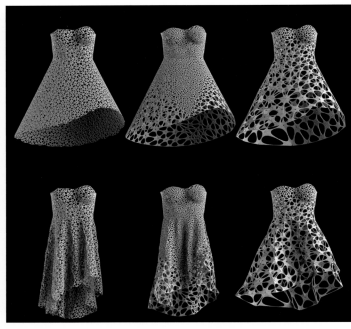

The scale of the tessellation can be adjusted to alter the drape of the dress

One of the most incredible potentials unlocked by 3D printing is the ability to fabricate an object made of articulated, movable parts in a single process. The complex tessellated or chain-link "fabrics" that can now be printed would require a very high amount of time and labor for assembly if produced using traditional means. This potential is currently being explored in depth by Nervous System, a Massachusetts-based generative design studio, who were originally presented with the challenge by Motorola's Advanced Technology and Projects division to design a viable system for the production of customized objects using personal 3D printers. The studio started investigating the possibility of printing flexible, jointed fabrics in compressed, "folded" arrangements and later "unfolding" them into larger shapes, like an entire dress, thus making the printing process more efficient and cheaper. Eventually, the project coalesced under the name Kinematics, a platform that is both a collection of jewelry made entirely of hinged triangular components and a custom software application for the design and modification of flexible jewelry. Nervous System has even created the Kinematics @ Home app for users with their own 3D printers, allowing them to download, modify, and print a bracelet file for free.

Tetra Kinematics 120-n necklace

Tetra Kinematics 175-n necklace

4

→ Appendix

Printing Things:
Visions and Essentials for 3D Printing

This book was conceived, edited, and designed by Gestalten.

Edited by Claire Warnier & Dries Verbruggen from Unfold, Sven Ehmann, Robert Klanten
Preface and features by Claire Warnier & Dries Verbruggen from Unfold
Project descriptions by Tamar Shafrir

Cover and layout by Floyd E. Schulze
Typeface: LL Replica by Norm

Proofreading by Wieners + Wieners
Printed by Nino Druck GmbH, Neustadt an der Weinstraße
Made in Germany

Published by Gestalten, Berlin 2014
ISBN 978-3-89955-516-5

For more information, please visit www.gestalten.com.

Bibliographic information published by the Deutsche Nationalbibliothek.
The Deutsche Nationalbibliothek lists this publication in the Deutsche Nationalbibliografie; detailed bibliographic data are available online at http://dnb.d-nb.de.

None of the content in this book was published in exchange for payment by commercial parties or designers; Gestalten selected all included work based solely on its artistic merit.

This book was printed on paper certified by the FSC®.

Gestalten is a climate-neutral company. We collaborate with the non-profit carbon offset provider myclimate (www.myclimate.org) to neutralize the company's carbon footprint produced through our worldwide business activities by investing in projects that reduce CO_2 emissions (www.gestalten.com/myclimate).